VOLUME SEVENTY TWO

Advances in
FOOD AND NUTRITION RESEARCH

Marine Carbohydrates: Fundamentals and Applications, Part A

ADVISORY BOARDS

KEN BUCKLE
University of New South Wales, Australia

MARY ELLEN CAMIRE
University of Maine, USA

ROGER CLEMENS
University of Southern California, USA

HILDEGARDE HEYMANN
University of California, Davis, USA

ROBERT HUTKINS
University of Nebraska, USA

RONALD JACKSON
Brock University, Canada

HUUB LELIEVELD
Global Harmonization Initiative, The Netherlands

DARYL B. LUND
University of Wisconsin, USA

CONNIE WEAVER
Purdue University, USA

RONALD WROLSTAD
Oregon State University, USA

SERIES EDITORS

GEORGE F. STEWART	(1948–1982)
EMIL M. MRAK	(1948–1987)
C. O. CHICHESTER	(1959–1988)
BERNARD S. SCHWEIGERT	(1984–1988)
JOHN E. KINSELLA	(1989–1993)
STEVE L. TAYLOR	(1995–2011)
JEYAKUMAR HENRY	(2011–)

VOLUME SEVENTY TWO

Advances in
FOOD AND NUTRITION RESEARCH

Marine Carbohydrates: Fundamentals and Applications, Part A

Edited by

SE-KWON KIM

*Department of Marine-bio Convergence Science,
Specialized Graduate School Science and
Technology Convergence,
Marine Bioprocess Research Center,
Pukyong National University,
Busan, South Korea*

AMSTERDAM • BOSTON • HEIDELBERG • LONDON
NEW YORK • OXFORD • PARIS • SAN DIEGO
SAN FRANCISCO • SINGAPORE • SYDNEY • TOKYO

Academic Press is an imprint of Elsevier

Academic Press is an imprint of Elsevier
225 Wyman Street, Waltham, MA 02451, USA
525 B Street, Suite 1800, San Diego, CA 92101-4495, USA
32 Jamestown Road, London, NW1 7BY, UK
The Boulevard, Langford Lane, Kidlington, Oxford, OX5 1GB, UK

First edition 2014

Copyright © 2014 Elsevier Inc. All rights reserved.

No part of this publication may be reproduced or transmitted in any form or by any means, electronic or mechanical, including photocopying, recording, or any information storage and retrieval system, without permission in writing from the publisher. Details on how to seek permission, further information about the Publisher's permissions policies and our arrangements with organizations such as the Copyright Clearance Center and the Copyright Licensing Agency, can be found at our website: www.elsevier.com/permissions.

This book and the individual contributions contained in it are protected under copyright by the Publisher (other than as may be noted herein).

Notices
Knowledge and best practice in this field are constantly changing. As new research and experience broaden our understanding, changes in research methods, professional practices, or medical treatment may become necessary.

Practitioners and researchers must always rely on their own experience and knowledge in evaluating and using any information, methods, compounds, or experiments described herein. In using such information or methods they should be mindful of their own safety and the safety of others, including parties for whom they have a professional responsibility.

To the fullest extent of the law, neither the Publisher nor the authors, contributors, or editors, assume any liability for any injury and/or damage to persons or property as a matter of products liability, negligence or otherwise, or from any use or operation of any methods, products, instructions, or ideas contained in the material herein.

ISBN: 978-0-12-800269-8
ISSN: 1043-4526

> For information on all Academic Press publications
> visit our website at store.elsevier.com

CONTENTS

Contributors	ix
Preface	xi

1. Isolation and Characterization of Chitin and Chitosan from Marine Origin — 1
Nitar Nwe, Tetsuya Furuike, and Hiroshi Tamura

1. Current Status of Chitin and Chitosan	2
2. Production of Chitin, Chitosan and Chito-oligosaccharide from Marine Materials	3
3. Physicochemical Properties of Chitin and Chitosan	9
References	14

2. Hybrid Carrageenans: Isolation, Chemical Structure, and Gel Properties — 17
Loic Hilliou

1. Introduction	18
2. Chemical Structure and Gel Mechanism	19
3. Isolation of Hybrid Carrageenan: From the Seaweeds to the Extracted Polysaccharide	23
4. Gel Properties	28
5. Perspectives	40
Acknowledgments	40
References	40

3. Isolation of Low-Molecular-Weight Heparin/Heparan Sulfate from Marine Sources — 45
Ramachandran Saravanan

1. Introduction	46
2. History of Heparin	47
3. Anticoagulant Activity of Heparin	48
4. Sources of Heparin	49
5. Difference Between Heparin and HS	52
6. Biomedical Significance of LMWH/HS	53
7. Isolation of LMWH Sulfate	54
8. Conclusion	58
References	59

4. Isolation and Characterization of Hyaluronic Acid from Marine Organisms — 61

Sadhasivam Giji and Muthuvel Arumugam

1. Introduction — 62
2. Targeted Sources for HA in the Past and Present Era — 62
3. Structure of Hyaluronic Acid — 64
4. Isolation Methods — 64
5. Characterization of Hyaluronic Acid — 67
6. Conclusion — 71
Acknowledgments — 73
References — 73

5. Extracellular Polysaccharides Produced by Marine Bacteria — 79

Panchanathan Manivasagan and Se-Kwon Kim

1. Introduction — 80
2. Extracellular Polysaccharides — 81
3. Roles of Microbial EPS in the Marine Environment — 81
4. Biosynthesis — 83
5. Source of Extracellular Polysaccharide-Producing Bacteria — 83
6. Isolation of Extracellular Polysaccharide-Producing Bacteria — 83
7. Marine EPS-Producing Microorganisms — 83
8. Biotechnological Applications of Extracellular Polysaccharides — 89
9. Conclusions — 90
Acknowledgments — 91
References — 91

6. Biological Activities of Alginate — 95

Mikinori Ueno and Tatsuya Oda

1. Introduction — 96
2. Macrophage-Stimulating Activities of Alginates — 98
3. Antioxidant Activities of Alginate — 106
References — 108

7. Biological Activities of Carrageenan — 113

Ratih Pangestuti and Se-Kwon Kim

1. Introduction — 113
2. Carrageenan Source and Extraction — 115
3. Biological Activities — 117

4. Food and Technological Applications of Carrageenan　121
　　5. Conclusions　122
　　References　123

8. Biological Activities of Heparan Sulfate　125
Muthuvel Arumugam and Sadhasivam Giji

　　1. Introduction　125
　　2. Materials and Methods　127
　　3. Results　129
　　4. Discussion　130
　　Acknowledgments　134
　　References　134

9. Beneficial Effects of Hyaluronic Acid　137
Prasad N. Sudha and Maximas H. Rose

　　1. Introduction　138
　　2. Structure of Hyaluronic Acid　139
　　3. Properties of Hyaluronic Acid　141
　　4. Modification of Hyaluronic Acid　141
　　5. Applications of Hyaluronic Acid　146
　　6. Conclusion　168
　　Acknowledgments　168
　　References　168

10. Fucoidans from Marine Algae as Potential Matrix Metalloproteinase Inhibitors　177
Noel Vinay Thomas and Se-Kwon Kim

　　1. Introduction　178
　　2. Sulfated Polysaccharides as Potential MMPIs　182
　　3. Conclusions and Further Prospects　188
　　Acknowledgments　189
　　References　189

11. Anticancer Effects of Fucoidan　195
Kalimuthu Senthilkumar and Se-Kwon Kim

　　1. Introduction　195
　　2. Seaweed Polysaccharides　196
　　3. Fucoidan and Cancer　200

	4. Conclusions	206
	Acknowledgment	207
	References	207

12. Anticancer Effects of Chitin and Chitosan Derivatives 215

Mustafa Zafer Karagozlu and Se-Kwon Kim

	1. Introduction	215
	2. Anticancer Activity as a Therapeutic Agent	217
	3. Anticancer Activity as a Carrier	220
	4. Conclusion	222
	References	222

Index 227

CONTRIBUTORS

Muthuvel Arumugam
Centre of Advanced Study in Marine Biology, Faculty of Marine Sciences, Annamalai University, Parangipettai, Tamil Nadu, India

Tetsuya Furuike
Faculty of Chemistry, Materials and Bioengineering, Kansai University, Suita, Osaka, Japan

Sadhasivam Giji
Centre of Advanced Study in Marine Biology, Faculty of Marine Sciences, Annamalai University, Parangipettai, Tamil Nadu, India

Loic Hilliou
Institute for Polymers and Composites/I3N, University of Minho, Guimarães, Portugal

Mustafa Zafer Karagozlu
Marine Bioprocess Research Center, Pukyong National University, Busan, South Korea

Se-Kwon Kim
Department of Marine-bio Convergence Science, Specialized Graduate School Science and Technology Convergence, Marine Bioprocess Research Center, Pukyong National University, Busan, South Korea

Panchanathan Manivasagan
Marine Biotechnology Laboratory, Department of Chemistry, Pukyong National University, Busan, South Korea

Nitar Nwe
Dukkha Life Science Laboratory, Thanlyin, Yangon, Myanmar; Faculty of Chemistry, Materials and Bioengineering, Kansai University, Suita, Osaka, Japan, and Bioprocess Technology, Food Engineering and Bioprocess Technology, Asian Institute of Technology, Klong Luang, Pathumthani, Bangkok, Thailand

Tatsuya Oda
Division of Biochemistry, Faculty of Fisheries, Nagasaki University, Nagasaki, Japan

Ratih Pangestuti
Research Center for Oceanography, The Indonesian Institute of Sciences, Jakarta, Republic of Indonesia

Maximas H. Rose
Department of Biology, Sri Sai Vidyasharam, Vellore, Tamil Nadu, India

Ramachandran Saravanan
Department of Marine Pharmacology, Faculty of Allied Health Sciences, Chettinad Hospital and Research Institute, Chennai, India

Kalimuthu Senthilkumar
Specialized Graduate School Science and Technology Convergence, Department of Marine Bio Convergence Science, Pukyong National University, Busan, and Department of

Nuclear Medicine, Kyungpook National University School of Medicine, Daegu, South Korea

Prasad N. Sudha
PG and Research Department of Chemistry, DKM College for Women, Thiruvalluvar University, Vellore, Tamil Nadu, India

Hiroshi Tamura
Faculty of Chemistry, Materials and Bioengineering, Kansai University, Suita, Osaka, Japan

Noel Vinay Thomas
Marine Biochemistry Laboratory, Department of Chemistry, Pukyong National University, Busan, South Korea

Mikinori Ueno
Division of Biochemistry, Faculty of Fisheries, Nagasaki University, Nagasaki, Japan

PREFACE

Carbohydrates are large molecules and divided into four main categories, monosaccharides, disaccharides, oligosaccharides, and polysaccharides. Polysaccharides from marine source have unique properties due to present extreme environment and widely used for food application. Biological and biomedical application of marine-derived polysaccharides has been explored in small amount. This book aims to collect the material for biological, biomedical, and industrial application of marine-derived polysaccharides.

This book has been divided into two volumes, each volume contains 12 chapters
- Chapters 1–5 provide the detailed information about isolation and characterization techniques of marine polysaccharides (chitin, chitosan, carrageenan, heparan sulfate, hyaluronic acid, and extracellular polysaccharides) in detail.
- Chapters 6–15 describe the usage of marine polysaccharides in biological applications such as matrix metalloproteinase inhibitory effect, anticancer, antiallergy, antioxidant, and antidiabetic effects.
- Chapters 16–18 deal about biomedical application of marine polysaccharides; tissue engineering, drug delivery, and gene delivery applications are explored well.
- Chapters 19 and 23 explain about the usage of marine polysaccharides in wastewater treatment and industrial application.
- Chapters 21, 22, and 24 deal about functional food, nutraceutical, pharmaceutical, and cosmeceutical value of marine polysaccharides.

This book provides cumulative information about marine polysaccharides and their biological, biomedical, and industrial applications. Hence, this book will be important reference for marine biotechnologist, natural product scientist, and whoever working on marine polysaccharides field.

CHAPTER ONE

Isolation and Characterization of Chitin and Chitosan from Marine Origin

Nitar Nwe*,†,‡,1, Tetsuya Furuike†, Hiroshi Tamura†,1

*Dukkha Life Science Laboratory, Thanlyin, Yangon, Myanmar
†Faculty of Chemistry, Materials and Bioengineering, Kansai University, Suita, Osaka, Japan
‡Bioprocess Technology, Food Engineering and Bioprocess Technology, Asian Institute of Technology, Klong Luang, Pathumthani, Bangkok, Thailand
1Corresponding authors: e-mail address: nitarnwe@yahoo.com; tamura@kansai-u.ac.jp

Contents

1. Current Status of Chitin and Chitosan 2
2. Production of Chitin, Chitosan and Chito-oligosaccharide from Marine Materials 3
   2.1 Production of chitin and chitosan 3
   2.2 Production of LMW chitosan and chito-oligosaccharides 6
3. Physicochemical Properties of Chitin and Chitosan 9
   3.1 Appearance of chitin and chitosan 9
   3.2 Turbidity of chitin and chitosan solution 11
   3.3 Determination of degree of deacetylation and degree of acetylation 11
   3.4 Determination of molecular weight of chitosan 13
References 14

Abstract

Nowadays, chitin and chitosan are produced from the shells of crabs and shrimps, and bone plate of squid in laboratory to industrial scale. Production of chitosan involved deproteinization, demineralization, and deacetylation. The characteristics of chitin and chitosan mainly depend on production processes and conditions. The characteristics of these biopolymers such as appearance of polymer, turbidity of polymer solution, degree of deacetylation, and molecular weight are of major importance on applications of these polymers. This chapter addresses the production processes and conditions to produce chitin, chitosan, and chito-oligosaccharide and methods for characterization of chitin, chitosan, and chito-oligosaccharide.

1. CURRENT STATUS OF CHITIN AND CHITOSAN

Chitin and chitosan are found as supporting materials in most of marine species such as shrimps, crabs, squids, lobsters, crayfishes, krill, and cuttlefishes. Among them, chitin and chitosan have been produced in laboratory and commercial scales from shells of shrimps and crabs and bone plates of squids and have been used to prepare various forms to apply as biomaterial matrixes in various products such as antioxidative agents, bone healing agent, cholesterol reduction agent, dental enamel de-remineralization, edible film, flocculating agent, growth-stimulating agent for plant, hemostatic agent, immune-stimulating agent, immobilization of cell, and juice clarification (Nwe, Furuike, Osaka, et al., 2011; Nwe, Furuike, & Tamura, 2011), in which physicochemical properties of chitin and chitosans such as molecular weight, degree of acetylation, and solubility are of major importance for success on those applications.

Nowadays, commercially, chitins and chitosans are produced from biowastes obtained from aquatic organisms, in which the physicochemical characteristics of chitosan are batch-to-batch variability due to seasonality of raw materials, freshness of the shell, quality of shell, species present, climate and distance of the shell supply, variability in raw materials, and difficulties in process control (Ashford et al., 1977 cited in Ornum, 1992; White, Farina, & Fulton, 1979). To solve these problems, several excellent research contributions on the production of chitin and chitosan from aquatic organisms and on the characteristics of chitin and chitosan have been made by scientists from all over the world for about 200 years. Nowadays, process for production of chitin and chitosan has been improved according to results obtained from research contributions.

All chitins and chitosans are not applicable in all sectors. For example, the application of chitosan as antibacterial agent or plant growth stimulator requires low-molecular-weight (LMW) chitosan. All chitosans do not support to make those applications successful. Up to now, the terms "chitin" and "chitosan" are used for all products of chitin and chitosan. There is a need to develop a nomenclature system to represent each type of chitin and chitosan product. In 2008, Prof. George A. F. Roberts from Nottingham University proposed to develop a nomenclature system for chitin and chitosan (Roberts, 2008 cited in Nwe, Furuike, Osaka, et al., 2011; Nwe, Furuike, & Tamura, 2011). He proposed that chitin and chitosan should be named with mole fraction of D-glucosamine or N-acetyl-D-glucosamine. In 2011 and 2013, we also proposed to develop a nomenclature system for chitin and chitosan (Nwe, Furuike, Osaka, et al., 2011; Nwe, Furuike, & Tamura, 2011; Nwe et al., 2013). To develop a systematic nomenclature system for chitin and chitosan, there is a need to establish international official standard methods to determine the degree of deacetylation (DD) and molecular weight of chitin and chitosan. International official standard methods to determine degree of deacetylation (DD) (ASTM F2260-03(2012)e1) and molecular weight (ASTM F2602-13) of chitin and chitosan have been published.

This chapter is the extended chapter of the previous chapters published in the book *Biodegradable Materials: Production, properties and applications* (2011), Chapter 2, Nova Science Publishers, Inc. and in the book *Marine Biomaterials: Characterization, Isolation, and applications* (2013), Chapter 4, CRC press, Taylor and Francis Group. This chapter presents the production processes of chitin, chitosan and chito-oligosaccharide from marine origin. In addition, methods for determination of particle size, turbidity, degree of deacetylation and degree of acetylation of chitin and chitosan, and of molecular weight of chitosan and chito-oligosaccharide (CTS-O) are presented.

2. PRODUCTION OF CHITIN, CHITOSAN AND CHITO-OLIGOSACCHARIDE FROM MARINE MATERIALS

The production of shrimp and crab meat in seafood industry results into massive amounts of shells of shrimps and crabs as biowaste (Fig. 1.1). This waste has been used as fertilizer and animal feed. Nowadays, this waste is used to produce chitin and chitosan in commercial scale. The production of chitin and chitosan from marine crustacean sources has been reported using several experimental conditions in laboratory scale.

2.1. Production of chitin and chitosan

The procedure for the production of chitin from shells of shrimps and crabs, and bone plates of squids includes demineralization and deproteination process and then chitosan is produced by deacetylation of chitin (Fig. 1.2). The different quality of chitosan from these sources was obtained by treatment of chitin with concentrated NaOH at different conditions (Table 1.1).

The demineralization of shrimp waste is completed within 15 min using 0.25 M HCl at room temperature (Roberts, 2008 cited in Nwe, Furuike, Osaka, et al., 2011; Nwe, Furuike, & Tamura, 2011). High degree of

Figure 1.1 Shells of shrimps obtained from seafood processing industry. *Photos were obtained from Great International Fisheries Ltd., Hlaing Thar Yar Industrial City, Zone (4), Yangon, Myanmar. Permission to use photos has been obtained from Dr. Nu Nu Win, Managing Director, Great International Fisheries Ltd., Hlaing Thar Yar Industrial City, Zone (4), Yangon, Myanmar.*

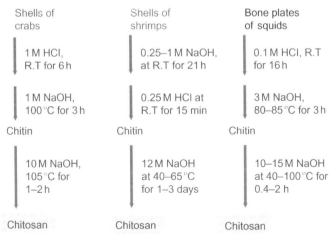

Figure 1.2 Production of chitosan from shells of crabs and shrimps and bone plates of squids (Lertsutthiwong, How, Chandrkrachang, & Stevens, 2002; Methacanon, Prasitsilp, Pothsree, & Pattaraarchachai, 2003; Roberts, 2008 cited in Nwe, Furuike, Osaka, et al., 2011; Nwe, Furuike, & Tamura, 2011; Tajik, Moradi, Rohani, Erfani, & Jalali, 2008; Trung, Thein-Han, Qui, Ng, & Stevens, 2006; Yen, Yang, & Mau, 2009). *(Reproduced from Nwe et al., 2013, Marine biomaterials: Characterization, isolation and applications, (pp. 45–60). CRC press, Taylor and Francis Group. Permission has been obtained.)*

Table 1.1 Condition for deacetylation of chitin and degree of N-deacetylation (DD) of chitosan obtained from different sources of chitin under various conditions

Source	Chitin/NaOH solution ratio (g/ml)	NaOH concentration (%)	Temperature (°C)	Time (min)	%DD (average)
Squid chitin	1:10	40	40	120	43.3[a]
		40	60	120	68[a]
		40	80	120	84[a]
		40	100	120	88.7[a]
		60	40	120	27.5[a]
		60	60	120	70.7[a]
		60	80	45	84.2[a]
		60	80	60	90.2[a]
		60	80	120	94.7[a]
		60	100	30	94.0[a]
		60	100	60	97.3[a]
Crab chitin	1:30	40	105	60	83.3[b]
				90	88.4[b]
				120	93.3[b]
Shrimp chitin	1:20	50	40	3 days	75.9[c]
Shrimp chitin	1:13	50	121	15	70[d]
Shrimp chitin	–	50	40	24 h	75[e]
Shrimp chitin	–	50	65	20 h	87[e]

[a]The degree of deacetylation (DD) was measured using ^{13}C CP/MAS NMR spectra (Methacanon et al., 2003).
[b]The degree of deacetylation (DD) was measured using infrared spectrometer (Yen et al., 2009).
[c]The degree of deacetylation (DD) was measured using HPLC (Lertsutthiwong et al., 2002).
[d]The degree of deacetylation (DD) was measured using titration method (Tajik et al., 2008).
[e]The degree of deacetylation (DD) was measured using 1st derivative UV spectrophotometry method (Trung et al., 2006).

deacetylation (80–100% DD) of chitosan could be obtained by treatment of chitin from squid pens with 10–15 M NaOH at 80–100 °C for 1–2 h (Table 1.1). The resultant chitosan has medium molecular weight. However, chitosans with medium molecular weight, 810 kDa, and various DD have been produced by treatment of chitin from shells of shrimps with

12.5 M NaOH under different treatment conditions (Trung et al., 2006), in which Lertsutthiwong et al. (2002) reported that the DD and molecular weight of chitosan produced from shells of shrimps depends on the conditions of extraction process, including the concentration of NaOH and HCl, the soaking time, and the sequence of treatments for deproteination, decalcification, and deacetylation. The scale-up shrimp chitin deacetylation was carried out with strong alkaline under ambient temperature for 3 days, and the degree of deacetylation was 75.9%. The higher degree of deacetylation of chitosan with 90% was obtained by second treatment with strong alkaline (Chinadit et al., 1998 cited in Nwe, Furuike, Osaka, et al., 2011; Nwe, Furuike, & Tamura, 2011). The molecular weight of native chitin is usually larger than 1 million, while commercial chitosan products fall between 100,000 and 1,200,000 (Muzzarelli, 1973 cited in Nwe, Furuike, Osaka, et al., 2011; Nwe, Furuike, & Tamura, 2011).

Moreover, the production of chitin and chitosan by chemical process has a lot of problems such as environmental pollution, inconsistent molecular weight, and degree of deacetylation. To solve these problems, fermentation processes have been studied to produce chitin and chitosan from shells of shrimps and crabs, and bone plates of squids (Nwe, Furuike, Osaka, et al., 2011; Nwe, Furuike, & Tamura, 2011). The deproteination of shrimp shell has been carried by fermentation with *Bacillus subtilis* (Sini, Santhosh, & Mathew, 2007). The enzymatic deacetylation of various chitin preparations (natural crystalline chitin, partially deacetylated chitin and reprecipitated chitin) has been investigated using crude chitin deacetylase (CDA) obtained from fermentation of *Rhizopus oryzae* growth in solid state and submerged fermentation medium (Fig. 1.3). The results showed that natural crystalline chitin is a very poor substrate for CDA (Aye, Karuppuswamy, Ahamed, & Stevens, 2006).

2.2. Production of LMW chitosan and chito-oligosaccharides

According to data of application of chitosan in plant tissue culture, LMW chitosans and CTS-O are more effective than high-molecular-weight (HMW) chitosan (Nwe, Furuike, & Tamura, 2010). "The LMW chitosan and CTS-O have been produced from the HMW chitosan by physical method such as treatment with gamma-ray irradiation and with microwave irradiation; chemical method such as treatment with dilute and concentrated HCl, with $NaNO_2$, with H_2O_2, with HNO_2, and with phosphoric acid; mechanical method such as treatment with sonication; and enzymatic

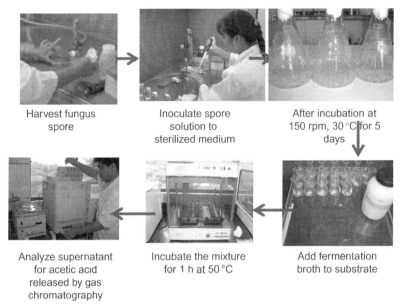

Figure 1.3 Production of crude chitin deacetylase and deacetylation of chitin using crude enzyme.

method such as treatment with cellulose, with lysozyme, with chitinase, with chitosanase, with lipase, with hyaluronidase, with papain, with pectinase, with pepsin, with protease and with hemicellulase (Nwe, Furuike, Osaka, et al., 2011; Nwe, Furuike, & Tamura, 2011). Although the different protocols for the production of low-molecular-weight chitosan and chito-oligosaccharide have been published, acid and enzymatic depolymerization of chitosan are most frequently used" (Nwe, Furuike, & Tamura, 2013).

2.2.1 Depolymerization of chitosan with chemical method

The chitosan with desire molecular weight can be obtained in depolymerization of chitosan with sodium nitrite by varying the reaction time (Luyen, 1994) and controlling the ratio of nitrite to chitosan (Thomas & Philip, 2000). Kasaai, Malaekeh, Charlet, and Arul (2001) reported that depolymerization of chitosan with HCl at 65 °C and with $NaNO_2$ at room temperature was more effective than depolymerization of chitosan with H_2O_2 at room temperature. The HNO_2 attacks the amino group of glucosamine units and cleavages the glycosidic linkage of the D–A and D–D which are known by determining the identity and the relative amounts of the new nonreducing ends of depolymerized chitosans (Vårum et al., 1996a cited in Vårum, Ottøy, Smidsrød, 2001). Thomas and Philip (2000) pointed out that sodium nitrite must be carefully removed from the LMW chitosan to use safely in medical application. Among the chemicals used, HCl is widely used for the production of LMW chitosan and CTS-O. The O-glycosidic linkages (depolymerization) and the N-acetyl linkages (de-N-acetylation) in the chitosan polymers are hydrolyzed during treatment with dilute and concentrated HCl (Vårum et al., 2001). The rate of hydrolysis of the glycosidic linkages was equal to that of N-acetyl linkage in dilute acid, in which the glycosidic linkages were hydrolyzed more than 10 times faster than the N-acetyl linkage in concentrated HCl (Vårum et al., 2001).

Under the treatment with dilute and concentrated HCl, nearly complete hydrolysis of chitosan (MW 10^4 Da) was observed using 6 M HCl at 110 °C for 13 h (Figs. 1.4 and 1.5). The chitosan samples, 30 mg, were digested with 6 ml of 6 M HCl at 110 °C for 10–13 h to obtain monomers of chitosan (Nwe, Furuike, Osaka, et al., 2011). Cabrera and Cutsem (2005) reported that treatment of chitosan (2 g) with 100 mL of concentrated HCl (37%) for 30 min at 72 °C under stirring was resulted in fragments with degrees of polymerization up to 16.

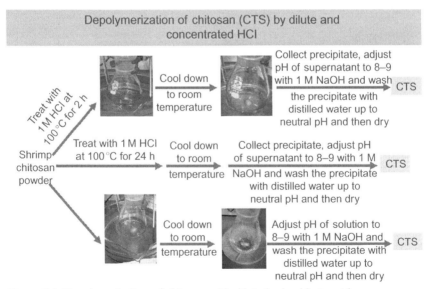

Figure 1.4 Depolymerization of chitosan with dilute hydrochloric acid.

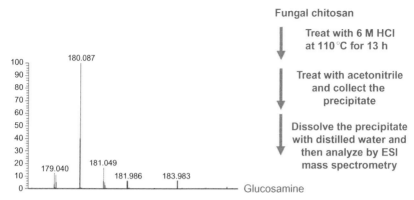

Figure 1.5 Depolymerization of chitosan with concentrated hydrochloric acid.

2.2.2 Depolymerization of chitosan with enzymatic method

Kasaai et al. (2001) reported that papain and wheat bran lipase were most effective for the depolymerization of chitosan as compared to other enzymes (pectinase, lysozyme, *Candida rugosa* lipase, and hyaluronidase). Chang, Liao, and Li (1998) reported that the degradation of chitosan to an average molecular weight of 30,000 Da, as observed during treatment with crude pepsin, was performed by the chitosanase present in the enzyme preparation and not by the proteolytic enzyme itself. The conditions for depolymerization of chitin and chitosan using some enzymes are shown in Table 1.2.

3. PHYSICOCHEMICAL PROPERTIES OF CHITIN AND CHITOSAN

3.1. Appearance of chitin and chitosan

Physical appearance testing is one of the most important tests performed on chitin and chitosan samples. This test carried out under solid state of the materials. The color of samples gives the information of purity of samples and information of production process. The test is carried out by viewing the material against a white background under laboratory lighting, and observation is recorded with the help of standard color chart.

The chitin and chitosan produced from shells of shrimp and crabs obtained chitin or chitosan flake. The fine powder of chitin and chitosan is produced using grinding and blending machine (Fig. 1.6). The distribution of particle size is determined by sieve analysis. Sieves are arranged on

Table 1.2 Enzymatic depolymerization of chitin and chitosan with various enzymes under various conditions

Enzyme	Chitin/chitosan	Treatment conditions pH	Temperature (°C)	Time (h)	Results
Pepsin[a]	Amorphous chitin (100 mg/20 mg enzyme)	5.4	44	24	71.5% of chitobiose, 19% of N-acetyl glucosamine, and 9.5% chitotriose
Neutral protease[b]	Chitosan (92% DD)	5.4	50	—	The degree of deacetylation of the main hydrolysis products decreased compared with the initial chitosan. The degree of polymerization of chito-oligomers was mainly from 3 to 8
Hemicellulase[c]	Chitosan (2.5%)	5.5	50	1–4	The enzymatic hydrolysis was endo-action and mainly occurred in a random fashion. The total degree of acetylation of chitosan did not change after degradation
Pectinex Ultra Spl[d]	Chitosan (82.9 kDa, DA 12%)	5.5	37	24	The enzymatic method yielded shorter fragments with a higher proportion of fully deacetylated chito-oligomers
Papain and Pronase[e]	Chitosan	3.5	37	1–5	Resulting in low molecular mass chitosan and chito-oligomeric-monomeric mixture

[a] Ilankovan, Hein, Ng, Trung, and Stevens (2006).
[b] Li et al. (2005).
[c] Qin et al. (2003).
[d] Cabrera and Cutsem (2005).
[e] Vishukumar, Varadaraj, Gowda, and Tharanathan (2005).
(Reproduced from Nwe et al., 2013, *Marine biomaterials: Characterization, isolation and applications*, (pp. 45–60). CRC press, Taylor and Francis Group. Permission has been obtained.)

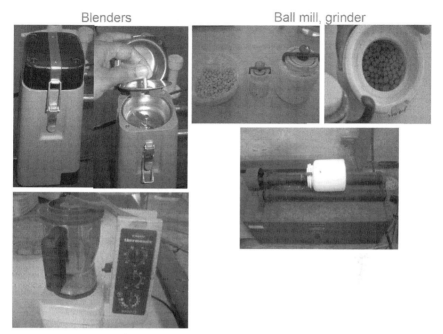

Figure 1.6 Grinding and blending machine.

shaker in a descending order with the sieve having the largest opening on top and a pan on the bottom. After that sample is poured into top sieve, shake the sieves for 5–10 min (Fig. 1.7).

3.2. Turbidity of chitin and chitosan solution

The turbidity of the chitosan solution was measured with a turbidimeter (Model 2100P portable turbidimeter, HACH Company, USA; Fig. 1.8).

3.3. Determination of degree of deacetylation and degree of acetylation

Many techniques have been developed for the determination of degree of acetylation of chitin and chitosan. Among them, liquid-state NMR spectroscopic method (Fig. 1.9), first-derivative UV spectrophotometric method, high-performance liquid chromatographic (HPLC) method, acid–base titration method, and PUV method are valid for perfectly soluble materials. For the elemental analysis, sample must be absolutely pure from residual proteins. "The X-ray diffraction, differential scanning calorimetry, and Fourier transform infrared spectroscopy (FTIR) (KBr pellet) methods showed poor

Figure 1.7 Sieves and mechanical shaker.

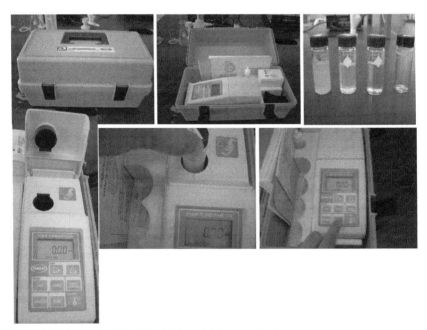

Figure 1.8 Determination of turbidity of chitosan solution.

Figure 1.9 Nuclear magnetic resonance (NMR) spectrometer.

accuracy with the samples of diverse preparations and sources" (Hein et al., 2008 cited in Nwe, Furuike, Osaka, et al., 2011; Nwe, Furuike, & Tamura, 2011).

FTIR methods are suitable for all chitosan samples (i.e., low to high DA chitosan samples); however, these methods depend on their calculation methods, sample forms, patterns of glucosamine, and N-acetyl glucosamine in chitosan molecules (Van de Velde & Kiekens, 2004 cited in Nwe, Furuike, Osaka, et al., 2011; Nwe, Furuike, & Tamura, 2011). The PUV, solid-state ^{13}C CP/MAS NMR, and acid hydrolysis–HPLC method showed the best methods to determine DA of chitin and chitosan; moreover, these methods are suitable for determination of acetyl content over the whole range of chitin and chitosan (Hein et al., 2008 cited in Nwe, Furuike, Osaka, et al., 2011; Nwe, Furuike, & Tamura, 2011). According to American Society for Testing and Materials organization, standard test method for determining degree of deacetylation in chitosan salts is proton nuclear magnetic resonance (^{1}H NMR) spectroscopy.

3.4. Determination of molecular weight of chitosan

For the determination of molecular weight of chitosan, gel permeation chromatographic (GPC), light-scattering, and viscosity methods are widely used. Chitosans produced using aquatic crustaceans under heterogeneous conditions and using fungal mycelia contain a population of chitosan molecules with varying molecular weight. Most research used the GPC method to determine the distribution of molecular weights of chitosan molecules in the sample. According to American Society for Testing and Materials organization, standard test method for determining the molar mass of chitosan and chitosan salts by size exclusion chromatography with multiangle light scattering detection.

REFERENCES

Aye, K. N., Karuppuswamy, R., Ahamed, T., & Stevens, W. F. (2006). Peripheral enzymatic deacetylation of chitin and reprecipitated chitin particles. *Bioresource Technology*, *97*, 577–582.

Cabrera, J. C., & Cutsem, P. V. (2005). Preparation of chitooligosaccharides with degree of polymerization higher than 6 by acid or enzymatic degradation of chitosan. *Biochemical Engineering Journal*, *25*, 165–172.

Chang, C. T., Liao, Y. M., & Li, S. J. (1998). Preparation of low molecular weight chitosan and chito-oligosaccharides by the enzymatic hydrolysis of chitosan. In R. H. Chen, & H. C. Chen (Eds.), *Advances in chitin science* (pp. 233–238). *Proceedings of the 3rd Asia Pacific symposium, Keelung*.

Ilankovan, P., Hein, S., Ng, C. H., Trung, T. S., & Stevens, W. F. (2006). Production of N-acetyl chitobiose from various chitin substrates using commercial enzymes. *Carbohydrate Polymers*, *63*, 245–250.

Kasaai, M. R., Malaekeh, M., Charlet, G., & Arul, J. (2001). Depolymerization of chitosan. In T. Uragami, K. Kurita, & T. Fukamizo (Eds.), *Chitin and chitosan, chitin and chitosan in life science: Proceedings of the Eighth International Chitin and Chitosan Conference and Fourth Asia Pacific Chitin and Chitosan Symposium* (pp. 28–35). Yamaguchi, Japan: Kodansha Scientific.

Lertsutthiwong, P., How, N. C., Chandrkrachang, S., & Stevens, W. F. (2002). Effect of chemical treatment on the characteristics of shrimp chitosan. *Journal of Metals, Materials and Minerals*, *12*, 11–18.

Li, J., Du, Y., Yang, J., Feng, T., Li, A., & Chen, P. (2005). Preparation and characterization of low molecular weight chitosan and chito-oligomers by a commercial enzyme. *Polymer Degradation and Stability*, *87*, 441–448.

Luyen, D. V. (1994). Partially hydrolyzed chitosan. In *Asia-Pacific chitin and chitosan symposium*, Malaysia: Bangi.

Methacanon, P., Prasitsilp, M., Pothsree, T., & Pattaraarchachai, J. (2003). Heterogeneous N-deacetylation of squid chitin in alkaline solution. *Carbohydrate Polymers*, *52*, 119–123.

Nwe, N., Furuike, T., Osaka, I., Fujimori, H., Kawasaki, H., Arakawa, R., et al. (2011). Laboratory scale production of ^{13}C labeled chitosan by fungi *Absidia coerulea* and *Gongronella butleri* grown in solid substrate and submerged fermentation. *Carbohydrate Polymers*, *84*, 743–750.

Nwe, N., Furuike, T., & Tamura, H. (2010). Production of fungal chitosan by enzymatic method and applications in plant tissue culture and tissue engineering: 11 years of our progress, present situation and future prospects. In M. Elnashar (Ed.), *Biopolymers* (pp. 135–162). Rijeka: SCIYO.

Nwe, N., Furuike, T., & Tamura, H. (2011). Chitosan from aquatic and terrestrial organisms and microorganisms: Production, properties and applications. In B. M. Johnson, &

Z. E. Berkel (Eds.), *Biodegradable materials: Production, properties and applications* (pp. 29–50). Hauppauge, New York: Nova Science Publishers.

Nwe, N., Furuike, T., & Tamura, H. (2013). Isolation and characterization of chitin and chitosan as potential biomaterials. In S. Kim (Ed.), *Marine biomaterials: Characterization, isolation and applications* (pp. 45–60). Boca Raton, FL United States: CRC Press, Taylor and Francis Group.

Ornum, J. (1992). Shrimp waste must it be wasted. *Infofish International, 6*, 48–52.

Qin, C., Du, Y., Zong, L., Zeng, F., Liu, Y., & Zhou, B. (2003). Effect of hemicellulase on the molecular weight and structure of chitosan. *Polymer Degradation and Stability, 80*, 435–441.

Sini, T. K., Santhosh, S., & Mathew, P. T. (2007). Study on the production of chitin and chitosan from shrimp shell by using *Bacillus subtilis* fermentation. *Carbohydrate Research, 342*, 2423–2429.

Tajik, H., Moradi, M., Rohani, S. M. R., Erfani, A. M., & Jalali, F. S. S. (2008). Preparation of chitosan from brine shrimp (*Artemia urmiana*) cyst shells and effects of different chemical processing sequences on the physicochemical and functional properties of the product. *Molecules, 13*, 1263–1274.

Thomas, P., & Philip, B. (2000). Production of partially degraded chitosan with desire molecular weight. In M. G. Peter, A. Domard, & R. A. A. Muzzarelli (Eds.), *Advance in chitin science* (pp 63–67). *Proceedings of the 3rd international conference of the European chitin society*, University of Potsdam.

Trung, T. S., Thein-Han, W. W., Qui, N. T., Ng, C. H., & Stevens, W. F. (2006). Functional characteristics of shrimp chitosan and its membranes as affected by the degree of deacetylation. *Bioresource Technology, 97*, 659–663.

Vårum, K. M., Ottøy, M. H., & Smidsrød, O. (2001). Acid hydrolysis of chitosans. *Carbohydrate Polymers, 46*, 89–98.

Vishukumar, A. B., Varadaraj, M. C., Gowda, L. R., & Tharanathan, R. N. (2005). Characterization of chito-oligosaccharides prepared by chitosanolysis with the aid of papain and Pronase, and their bactericidal action against *Bacillus cereus* and *Escherichia coli*. *Biochemical Journal, 391*, 167–175.

White, S. A., Farina, P. R., & Fulton, I. (1979). Production and isolation of chitosan from *Mucor rouxii*. *Applied and Environmental Microbiology, 38*, 323–328.

Yen, M., Yang, J., & Mau, J. (2009). Physicochemical characterization of chitin and chitosan from crab shells. *Carbohydrate Polymers, 75*, 15–21.

CHAPTER TWO

Hybrid Carrageenans: Isolation, Chemical Structure, and Gel Properties

Loic Hilliou[1]

Institute for Polymers and Composites/I3N, University of Minho, Guimarães, Portugal
[1]Corresponding author: e-mail address: loic@dep.uminho.pt

Contents

1. Introduction — 18
2. Chemical Structure and Gel Mechanism — 19
   2.1 Seaweeds chemistry — 19
   2.2 Hybrid carrageenan macromolecular structure — 21
3. Isolation of Hybrid Carrageenan: From the Seaweeds to the Extracted Polysaccharide — 23
   3.1 Season variability and postharvest storage of algal material — 24
   3.2 Alkali pretreatment of seaweeds to tune the hybrid carrageenan chemistry — 26
   3.3 Aquaculture in the dark: An eco-friendly alternative to alkali treatment of carrageenophytes? — 26
4. Gel Properties — 28
   4.1 Rotational rheometry — 28
   4.2 Experimental considerations on the rheology of carrageenan gels — 30
   4.3 Penetration tests — 33
   4.4 Effects of salt type and concentration — 33
   4.5 Effects of hybrid carrageenan concentration — 36
   4.6 Relationships between the chemical structure and the gel properties — 38
   4.7 Large deformation behavior and gels under steady flow — 39
5. Perspectives — 40
Acknowledgments — 40
References — 40

Abstract

Hybrid carrageenan is a special class of carrageenan with niche application in food industry. This polysaccharide is extracted from specific species of seaweeds belonging to the Gigartinales order. This chapter focuses on hybrid carrageenan showing the ability to form gels in water, which is known in the food industry as weak kappa or kappa-2 carrageenan. After introducing the general chemical structure defining hybrid carrageenan, the isolation of the polysaccharide will be discussed focusing on the interplay between seaweed species, extraction parameters, and the hybrid carrageenan chemistry. Then, the rheological experiments used to determine the small and large deformation behavior of gels will be detailed before reviewing the relationships between gel properties and hybrid carrageenan chemistry.

1. INTRODUCTION

Carrageenans are natural polymers contained in specific species of red seaweeds belonging to the Gigartinales order. These are polysaccharides showing a variety of chemical structures, resulting from a complex interplay between the seaweeds species, the seaweed life stage, and the extraction process used to recover the polysaccharide. Among the various types of carrageenans showing different gelling or viscosity enhancement properties in aqueous solutions, hybrid carrageenans have recently received increased interest (van de Velde, 2008). The latter is motivated by the steadily increasing demand for gelling additives for food and nonfood application, which puts under pressure the farming of seaweeds producing kappa-carrageenan (K) and iota-carrageenan (I) (Bixler, 1996; Bixler & Porse, 2011). Thus, alternative algal resources for carrageenan production are highly demanded (Bixler & Porse, 2011; McHugh, 2003), and seaweeds producing hybrid carrageenans can be a solution to the issue triggered by the market. Recently, hybrid carrageenans were found to positively replace mixtures of K and I used in niche application in dairy food (Bixler, Johndro, & Falshaw, 2001; Villanueva, Mendoza, Rodrigueza, Romero, & Montaño, 2004). In spite of the industrial need and interest in using hybrid carrageenans, there is a lack in the literature for the structural and mechanical characterization of hybrid carrageenan gels (van de Velde, 2008), which explains why the relationships between the hybrid carrageenan chemical structure, the gel microstructure, and the gel mechanical properties are not yet understood.

This chapter focuses on the gel properties of hybrid carrageenan and addresses the relationships identified between the seaweeds biology, the extraction parameters, the chemical structure, and the gel properties. First, the biology and chemical composition of seaweeds producing hybrid carrageenan will be described. Then, the chemical structure of gelling hybrid carrageenan will be introduced together with the general proposed mechanisms for gel formation for K and I in the presence of salt. As this chapter is concerned with gelling hybrid carrageenan, two types of polysaccharides are solely discussed, namely, kappa/iota-hybrid carrageenan (KI) and their biological precursor kappa/iota/mu/nu-hybrid carrageenan (KIMN). Thus, the gel formation builds up on the mechanism devised for K and I. Then, the extraction process used to isolate the polysaccharide and the effect of extraction parameters on the macromolecular structure and chemistry of recovered polysaccharides will be discussed. With the characterized hybrid carrageenan in hand, gels will be formed and the rheological technique used to analyze their mechanical properties will be introduced before describing the effects of salt and polysaccharide concentrations on gel setting and elastic properties. Then, the interplay between hybrid carrageenan chemical structure and the gel properties will be discussed. Finally, the effect of steady flow on the gel setting and gel properties will be addressed as its relevance to the delivery of new food ingredients such as fluid gels or microgels (Garrec & Norton, 2012) and to the industrial processing of carrageenan is bright.

2. CHEMICAL STRUCTURE AND GEL MECHANISM
2.1. Seaweeds chemistry

Before tackling the chemical structure of KI and KIMN, it is imperative to look at the chemical composition of seaweeds belonging to the Gigartinaceae, Petrocelidaceae, and Phylophoraceae families which are the major carrageenophytes used for the production of gelling hybrid carrageenan (see for instance the Stancioff's diagram in Bixler, 1996). Fourier transform infrared diffuse reflectance spectroscopy (DRIFT) is a versatile spectroscopic method which boosted the chemical analysis of seaweeds as no sample preparation but grinding dried seaweeds is required. DRIFT was applied to screen for the chemical composition of Gigartinales by Chopin, Kerin, and Mazerolle (1999). This extensive study performed on more than 50 species of the Gigartinales order confirmed that the chemical composition of seaweeds depends on its life stage, vegetative, or reproductive and in the latter case depends also on the gender of the gametophyte of specific seaweeds. Vegetative Gigartinales seaweeds are made of highly sulfated and nongelling carrageenan of the lambda-type, whereas the reproductive life stage produces K, I mu-carrageenan (M) and nu-carrageenan (N) in fronds and thalli of female and male gametophytes. The disaccharide units corresponding to these carrageenans are displayed in Fig. 2.1. This general picture was reached in earlier studies which relied on the isolation of polysaccharides with the inherent polymer chemical modification associated with the extraction process (see the tables in Chopin et al., 1999 where a direct comparison between DRIFT results and reports from the literature is offered). The impact of such modification on the qualitative chemical

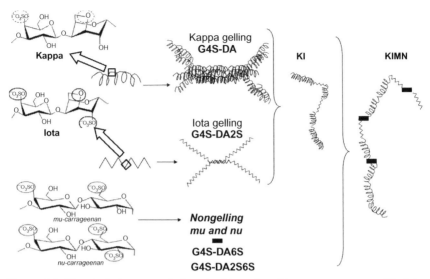

Figure 2.1 Chemical structure of disaccharide units of kappa-carrageenan (K), iota-carrageenan (I) mu-carrageenan (M), and nu-carrageenan (N) which are the building blocks of the gelling hybrid carrageenan KI and KIMN. The gelling mechanisms for K and I are depicted together with the block copolymer structure of KI and KIMN.

analysis of compounds contained in the seaweed was pointed recently in a study on *Mastocarpus stellatus* (Azevedo et al., 2013). DRIFT spectra of the native extracts obtained without alkali treatment showed all characteristic bands of K, I, M, and N disaccharide units, whereas both fronds and thalli did not show the specific band assigned to I, showing up at 805 cm^{-1}. The relative content in K, I, M, and N is however difficult to assess with this semiquantitative spectroscopic technique and thus hardly shows that the ratio between K and I is specific to each seaweed. This is illustrated in Fig. 2.2 where the DRIFT spectra of *M. stellatus* and *Chondrus crispus* hand collected on the Northern Portuguese coast are displayed, together with the spectra of commercial K and I (Sigma-Aldrich, Germany). Fronds and thalli of seaweeds were scratched on the DRIFT accessory pad of an FTIR spectrometer (Spectrum 100, PerkinElmer Ltd., UK), whereas powders were directly laid on the accessory. Both *M. stellatus* and *C. crispus* show the diagnose bands for I (805 cm^{-1}) and K (930 and 845 cm^{-1}). However, assessing whether *C. crispus* contains more K than *M. stellatus* is hard since ratios of K over I band intensities are 1.4 ± 0.2 for *C. crispus* against 1.1 ± 0.3 for *M. stellatus*, with errors computed from the averaging of five replicates from different fronds. Thus, one relies on extracting the polysaccharide from the

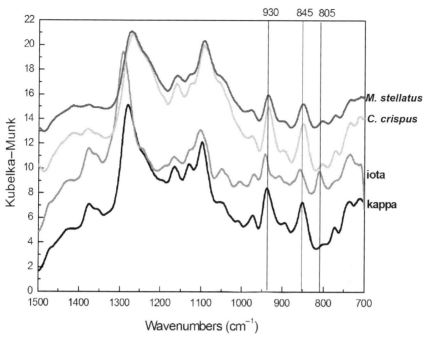

Figure 2.2 DRIFT spectra of Gigartinales and of commercial carrageenan. From bottom to top: kappa-carrageenan, iota-carrageenan, *Condrus crispus* (frond of a female gametophyte), and *Mastocarpus stellatus* (frond of a female gametophyte). Vertical lines indicate the bands assigned to galactose (975 cm^{-1}), 3,6-anhydrogalactose—DA (930 cm^{-1}), the sulfate group on the fourth carbon of the galactose—G4S (845 cm^{-1}), and the sulfate group on the second carbon of the 3,6-anhydrogalactose—DA2S (805 cm^{-1}). The latter band is specific to iota-carrageenan.

seaweed and analyzing by ^{1}H NMR the KI which are soluble in water to compute such ratios. A nice example of such exercise can be found in van de Velde et al. (2005) where three seaweed species harvested on the Portuguese coast showed K over I ratios ranging from 0.02 to 1.

2.2. Hybrid carrageenan macromolecular structure

The chemical structure of hybrid carrageenan has been vividly debated between two schools, and it is the author's opinion that the issue has been only recently solved by the publication of two critical papers (Guibet et al., 2008; van de Velde, Peppelman, Rollema, & Tromp, 2001). In contrast to gelling ideal K and I which are essentially homopolymers of kappa-carrageenan and iota-carrageenan disaccharide units, KI are copolymers made of blocks of K and I with various lengths and random distribution in the chain, whereas KIMN possess blocks of K and I separated by disaccharide units of mu-carrageenan and nu-carrageenan (Guibet et al., 2008). The chemical structures of K, I, KI, and KIMN are depicted in Fig. 2.1. K and I are ideally (because nature is not perfect, even commercial seaweeds producing almost pure K will biosynthesize a KI with 3–10 mol% of I, see for instance Table 2.1 in van de Velde, 2008) linear polymers which are built on an alternating β-1,3-linked D-galactopyranosyl (**G**) and α-1,4-linked D-galactopyranosyl (**D**) residues. Several types of carrageenans are recognized according to sulfate (**S**) substitution of hydroxyl group/s in various positions in the disaccharide backbone and cyclization of the **D** units to form an anhydro ring (**A**). Hence, K is chemically **G4S-DA**, I is **G4S-DA2S**. Then, M and N are more sulfated than K and I, respectively, bearing an additional **S** in place of the **A**. Thus, M is chemically **G4S-D6S** and N is chemically **G4S-D2S6S**. The copolymer structure of KI and KIMN stems from the fact that a commercial alkali extracted KI free of any contaminant, such as Floridean starch or pyruvate, cannot be separated into K and I by fractionation in KCl (van de Velde et al., 2001). Similar results were obtained earlier with commercial alkali-treated extracts from *C. crispus* and *M. stellatus*, though the authors did not conclude about a copolymer structure (Rochas, Rinaudo, & Landry, 1989). Under certain temperature, polysaccharide and KCl concentrations, K form gels whereas I does not. Thus, in a mixture of K and I, I can be separated from the K gel phase by centrifugation. In a recent study where the phase diagrams of an alkali extracted KI from hand collected *M. stellatus* were built in NaCl and KCl without relying on centrifugation, no phase separation between a gel and a liquid phase could be found with KCl in contrast to an equivalent mixture of K and I (Azevedo, Bernardo, & Hilliou, 2014). Indeed, the differences between the physical properties of mixtures of K plus I and KI reported long ago by Stancioff (1981) prompted this author to conjecture

Table 2.1 Power law exponents for the scaling of G_0 measured with different cooling rates and for the scaling of the apparent Young modulus E_{app} with the carrageenan concentration in 0.05 M KCl

Cooling rate	E_{app}		G_0		
	KI	K+I	KIMN	I	K
5 °C/min	1.37 ± 0.08	3.0 ± 0.2	3.18 ± 0.05	1.76 ± 0.05	2.08 ± 0.13
1 °C/min	–	–	3.11 ± 0.1	2.03 ± 0.09	2.92 ± 0.06

K + I stands from a mixture of commercial K and I.

a copolymer structure for KI instead of a mixture of two carrageenan homopolymers, though the exact structure of mixtures of K and I gels remained unclear. Once the copolymer structure is established, one might ask whether the polysaccharide chain is made of a statistical distribution of K and I disaccharide units, or a blocky arrangement of alternating K and I segments, and in this case, how long are these blocks. Guibet et al. (2008) used enzymatic degradation of KI and analyzed the degradation products. They showed that KI were made of blocks of K disaccharide units, of sequences enriched in K disaccharide, and of sequences enriched in I disaccharide units. But how long are the blocks and sequences of enriched I and K disaccharide units? We may build on the fact that KI form gels. For this, we need to recall here the currently accepted mechanism for gel formation in carrageenan (Rees, Morris, & Robinson, 1980; Viebke, Piculell, & Nilsson, 1994) which is schematically illustrated in Fig. 2.1. The gel formation is essentially driven by phase separation. Upon cooling down a hot carrageenan solution in the presence of salt, the polysaccharide chains undergo a coil-to-helix transition. With further cooling and time, helices aggregate and form domains pervading a three-dimensional network. The exact helix type, aggregation scheme, and role of salt are still an unresolved issue (Piculell, 1995). However, K gels show large thermal hysteresis between gel setting and melting, suggesting a superhelical gel structure (Viebke et al., 1994). In contrast to this, I gels show virtually no thermal hysteresis, indicating less helix aggregation or networking at the helical level; that is, cross-links in the network are actually made of double helices (Rees, 1972). Since KI can form gels, it is clear that at least one type of block is long enough to allow for the coil-to-helix conformational transition and subsequent aggregation. Thus, building on the random length distribution model fitted to high-sensitivity differential scanning calorimetry and optical rotation data of KI (van de Velde et al., 2005), it has been shown that K blocks have a minimum length between 14 and 8 disaccharide units, whereas I blocks show minimum lengths between 5 and 2 disaccharide units.

3. ISOLATION OF HYBRID CARRAGEENAN: FROM THE SEAWEEDS TO THE EXTRACTED POLYSACCHARIDE

Gelling carrageenans K and I are soluble in hot water. Therefore, the KI extraction is pretty much similar to the extraction route employed for commercial K and I (McHugh, 2003). Basically, algal material is suspended in hot water during a certain amount of time and at a defined temperature.

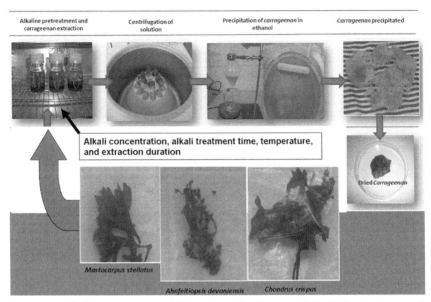

Figure 2.3 Alkali pretreatment of *M. stellatus*, *C. crispus*, and *Ahnfeltiopsis devoniensis* and hot extraction of hybrid carrageenan. Parameters affecting the polysaccharide chemical structure and molecular mass are grouped in a box.

Then, the solid algal material is separated from the sol phase which is a water solution of KI, and the polysaccharide is eventually recovered by precipitation in alcohol. In this way, KIMN are recovered. Alkali conversion of M and N into K and I, respectively, is performed either directly on the isolated KIMN. Alternatively, seaweeds are pretreated with alkali prior to extraction which is the preferred route in the carrageenan industry. The general extraction process is illustrated in Fig. 2.3.

3.1. Season variability and postharvest storage of algal material

As indicated earlier, the biological life cycle and alternation of genders in Gigartinales lead to significant variation in the chemistry of extracted carrageenans. Algal growth and aging, which are conditioned by environmental factors, have also an impact on the carrageenan chemical structure and yield. M and N are biological precursors of K and I, respectively. They are produced in the outer mucilageous layer of algal cell walls during the seaweed growth and are later transformed into K and I through biosynthetic pathways which still need to be elucidated for some Gigartinales (Fogg, 1964; McCandless, 1981). Extracting carrageenan from vegetative or reproducing seaweeds as well as from growing seaweeds or from mature seaweeds should give polysaccharides with variation in the chemical composition. A report on *Kappaphycus striatum* showed that young seaweeds produce more sulfated carrageenan than mature plants, suggesting that the carrageenan biosynthesis is related to algal growth (Mendoza, Ganzon-Fortes, Villanueva, Romero, & Montaño, 2006). Thus, K over I ratios varying from 0.2 to 0.54 in KI extracted from five *Ahnfeltiopsis devoniensis* samples come as no surprise (see Table 2.1 in van de Velde et al., 2005), since such seasonal variability has been observed in other Gigartinales (Freile-Pelegrin & Robledo, 2006; Pereira & Mesquita, 2003, 2004; Wakibia, Bolton, Keats, & Raitt, 2006). However, correlations between KI chemical variation and seaweeds variation in populations of fructified or nonfructified female and male gametophytes as well as sporophytes are still scarce (Amimi, Mouradi, Bennasser, & Givernaud, 2007; Hilliou et al., 2012), as reports rather focus on the impact of seasonal variations on carrageenan yield and gel quality (Freile-Pelegrin & Robledo, 2006; Fuller & Mathieson, 1972; Wakibia et al., 2006). Seasonal variation in the chemical structure of KI extracted from *M. stellatus* harvested on the Northern Portuguese coast was recently revealed (Hilliou et al., 2012). The chemical structure of KI assessed by both FTIR and ^{1}H NMR was also compared to the populations of fructified and nonfructified gametophytes. Note that for *M. stellatus*, the vegetative plants, the sporophytes, form a crust on the rocks which can hardly be removed for harvesting. This comprehensive study also included the rheological characterization of gels made from native and alkali extracted KI. Results showed seasonal variation in the sulfate degree of native KI, which was correlated with nonfructified gametophytes. As for the alkali extracted KI, seasonal variation in N and K was evidenced, but no correlation with the biology of *M. stellatus* was found. Maximum population of fructified fronds was found in summer (August) when seaweeds growth is maximum. Accordingly, more sulfated native KI were extracted in August and the content in N was the highest.

The impact of seaweed postharvest storage on the chemical–physical properties of hybrid carrageenan has virtually not been studied, in contrast to agar (Romero, Villanueva, & Montano, 2008). This topic has been recently revisited for *M. Stellatus*. Thirty-six month storage of dried seaweeds (48 h in an oven with air convection at 60 °C to achieve a water content below 8 wt%) in sealed black plastic bags did not affect significantly the chemical structure and the functional viscoelastic properties of alkali extracted KI (Hilliou et al., 2012).

3.2. Alkali pretreatment of seaweeds to tune the hybrid carrageenan chemistry

The alternative chemical route to the biosynthetic pathway converting carrageenan bioprecursors into less sulfated carrageenan presents various parameters which need to be optimized with respect to a targeted functional property for the hybrid carrageenan. A systematic variation of the type and concentration of alkali, of the time of alkali treatment, of the pH, temperature, and time of the extraction has been conducted with a model carrageenophyte in order to assess the effects of these parameters on the final gel properties and chemical structures of recovered KI (Azevedo et al., 2013; Hilliou, Larotonda, Abreu, et al., 2006; Hilliou, Larotonda, Sereno, & Gonçalves, 2006). The seaweed chosen for this study was *M. stellatus* as only gametophytes can be hand collected, in contrast to, for example, *C. crispus*, thus avoiding the tedious separation from vegetative thalli motivated by the reasons given above. KI and KIMN with different K over I ratios and relative contents in M and N could be isolated, thus giving access to the delivery of a palette of hydrocolloids with different gel properties, ranging from the weak elasticity of I gels to the much stiffer elasticity of K gels. Prolonged Na_2CO_3 or KOH pretreatments favor the conversion of more M into K (Azevedo et al., 2013; Hilliou, Larotonda, Abreu, et al., 2006; Hilliou, Larotonda, Sereno, et al., 2006), whereas treatments using NaOH seem to efficiently convert M and N from the first hour. For each type of alkali used, an optimum concentration allows the recovery of KI or KIMN (with Na_2CO_3) with best gel elasticity. The study with NaOH and KOH also confirmed that beyond the optimum concentration and time, KI are degraded resulting in a significant loss in molecular mass while keeping the same chemistry. Preliminary results with *C. crispus* and *Ahnfeltiopsis devoniensis* suggest that the optimum alkali treatment parameters (duration and alkali concentration) toward best KI chemistry with minimum M and N contents and no loss in molecular mass (these are the KI characteristics which likely will allow for best gelling properties) are specific to the seaweed.

3.3. Aquaculture in the dark: An eco-friendly alternative to alkali treatment of carrageenophytes?

Alkali soaking of seaweeds at hot temperature at the industrial scale generates effluents which need treatment prior to discharge and is energetically demanding. Inspired by earlier studies with agarophytes, Villanueva, Hilliou, and Sousa-Pinto (2009) explored the effect of dark cultivation of

C. crispus prior to KI extraction on the chemical and gel properties of the hybrid carrageenan. Dark treatment was previously applied to seaweeds of the Gracilariaceae family for agar production from 1990 till now, but with variable quality results which reveal that a definitive substitution of the alkali treatment by the "dark treatment" has not been achieved. After 10 days cultivation of wild-harvested seaweeds in the dark, KI were extracted which showed K over I ratios similar to the one computed from KI extracted after KOH treatment of wild seaweeds, that is, 69/31 (% mol/% mol). In addition, the extraction from dark-treated seaweeds yielded more KI than the extraction of alkali-treated *C. crispus*, thus suggesting that the eco-friendly alternative is actually cost effective. Contrary to the results for *C. crispus*, dark treatment did not improve the carrageenan gel quality of *M. stellatus*. FTIR analysis showed no appreciable decrease in sulfate and increase in 3,6-anhydrogalactose contents of extract after the dark treatment, nor after alkali treatment. ^{1}H NMR spectra of KI extracted from the harvested seaweeds and from the seaweeds cultivated in the dark do not show qualitative or quantitative differences: all three spectra in Fig. 2.4 show the signals assigned

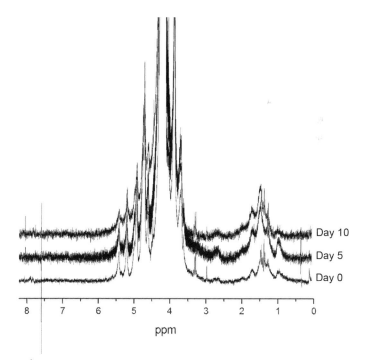

Figure 2.4 ^{1}H NMR spectra of native KI extracted with no alkali from *M. stellatus* freshly harvested (day 0), and cultivated 5 days (day 5) and 10 days (day 10) in the dark.

to K (5.11 ppm) and to I (5.31) in agreement with earlier results on native KI obtained from *M. stellatus* (Azevedo et al., 2013), and no difference in signal intensities is detected. This indicates that carrageenan precursors M and N (with sulfate ester at 6-position of 4-linked galactose units) production in the seaweed material is nil, probably due the low growth rate of the seaweed in culture, even in the presence of light. Thus, such precursors cannot be modified in K and I during cultivation in the dark or by alkali treatment. This suggests that the seaweed material hand collected for this study was at its mature stage and growth. As discussed above, this implies that carrageenan biosynthesis was minimal. Indeed, seaweeds were collected in December 2007 and usually no growth occurs during this winter period. Thus, such experiment should be reproduced with a different algal material before concluding about the effectiveness of dark treatment for *M. stellatus*. Also, these results point to the fact that dark cultivation should be tested for each carrageenophyte producing KI as no general trend is expected since each species shows different contents in biological precursors (see table 2.1 in van de Velde et al., 2005).

4. GEL PROPERTIES

Before reviewing the gel properties of KI, it is essential to recall here the basic principles of rheology and rheometry, whereas more tutorial can be found elsewhere (see for instance Barnes, Hutton, & Walters, 1989 for an introductory text book on rheology). Obviously, the focus here will be on the rheological characterization of gels.

4.1. Rotational rheometry
4.1.1 Small deformation
Small amplitude oscillatory shear (SAOS) experiments are the method of choice to assess the mechanical spectra of viscoelastic material. Mechanical spectra give information about the intrinsic viscoelastic properties over a wide range of frequencies, thus evidencing possible relaxation processes which are inherently related to equilibrium structural properties of the quiescent material. It is thus critical to study samples which are not evolving with time, as the record of spectra down to frequencies as small as 0.001 Hz can take more than 1000 s. The principle of SAOS is illustrated in Fig. 2.5. The sample is sandwiched between the two disks of a rotational rheometer. It is more common to find rheological data obtained with a stress-controlled rheometer, as compared to a strain-controlled rheometer,

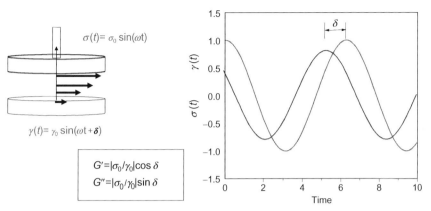

Figure 2.5 Principle of a small amplitude oscillatory shear test.

as the former is less expensive. In this case, the top disk will apply a sinusoidal stress of amplitude σ_0 and frequency ω, generated by a magnetic drive (see Barnes & Bell, 2003 for an historical perspective on stress rheometers). The bottom disk is fixed and does not move. Actually, in the cheapest versions, this disk is a Peltier plate which allows the temperature control of the sample. The sample deforms as a result of the imposed stress. This deformation is transmitted to the upper disk whose deflection is read from an optical encoder. The stress amplitude should be kept small enough in order to maintain the sample at equilibrium, that is, not induce orientation, stretching, breaking, or other phenomena which will induce a nonlinearity between shear stress and shear strain, resulting in the record of a nonsinusoidal deformation. Indeed, the rheometer compares the amplitudes of both sinusoidal stress σ_0 and sinusoidal deformation γ_0, as well as the phase shift δ between the two sinusoidal signals. Thus, it is critical to compare two sinusoidal signals to output a correct phase shift; otherwise, most of the phase shift information will be contained in the harmonics of the periodical strain signal (see below). From stress and strain amplitudes and phase shift, the storage modulus G' and the loss modulus G'' are computed to describe the elastic and viscous characters, respectively, of the sample for the input frequency. Equations defining G' and G'' are given in Fig. 2.5. By sweeping the frequency, the rheometer records the frequency dependence of both G' and G'', thus giving the mechanical spectrum of the sample which is essentially the mechanical picture of the sample. Again, it is essential to have a quiescent sample between the disks to avoid taking a blurred picture. An additional SAOS test includes fixing both stress and frequency and sweeps the temperature to measure the strain response. This temperature sweep test is used to determine a solid-to-liquid transition temperature, or melting or gelling transitions, by picking up the temperature at which G' crosses over G'', that is, the temperature at which the sample is as much liquid as solid at the probing frequency. Examples of mechanical spectra and temperature sweeps are given in Fig. 2.6.

4.1.2 Large deformation: Nonlinear rheology

When deformation or stress is large and fast enough, a gel is forced to stretch or break, whereas a gelling solution will be forced to flow and the structure will align in a specific direction or rearrange in a new structure. Thus, the rheological response will depend on the magnitude of stress and deformation, and a nonlinear regime between stress and strain is found. In oscillatory shear, this means that the strain response to large amplitude sinusoidal shear (LAOS) stress is no more sinusoidal, and the mathematics for computing the viscous and elastic contributions to the material response needs to consider the harmonic decomposition of the periodic strain signal. Different methods can be adopted, and this topic has been reviewed recently (Hyun et al., 2011). Fourier transform rheology (FTR) is one of these methods and its principle is illustrated in Fig. 2.7. In gels, FTR essentially allows for the characterization of strain-hardening behavior ($\phi_{3w1} - 3\phi_{w1}$ should be 0, and I_{3w1}/I_{w1} shows a quadratic increase with the strain) and of the critical strain for irreversible break up γ_C (even harmonics should show up as a result of loss of shear symmetry and reversibility). The concentration scaling of these parameters can be rationalized by theoretical models to relate the LAOS characteristic to gel structural parameters (Hilliou, Wilhelm, Yamanoi, & Gonçalves, 2009).

4.2. Experimental considerations on the rheology of carrageenan gels

Preparing KI or KIMN gels and transferring them to a rheometer is not the method of choice. First, because these gels are too soft or brittle to support the handling and loading between the disks. Second, because a static compression should be applied in the gel to ensure for good stress and strain transmission and limit slip problems. A static compression has a negative effect on the rheological results since both G' and G'' will increase proportionally (in case of small compression) or not with the compressive stress. Alternatively, one may choose to form the gel *in situ*, that is, by loading a hot carrageenan solution between the disks and cool down until gel formation. Then,

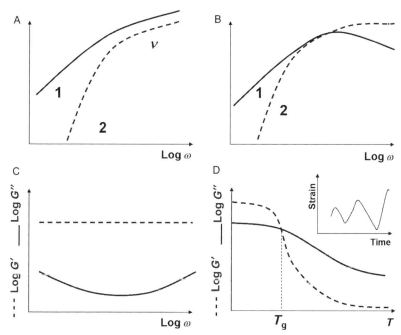

Figure 2.6 Mechanical spectra of a dilute polymer solution (A), of a semidilute entangled polymer solution (B), and of a polymer gel (C), and temperature sweep carried out for characterizing the gelling behavior during the cooling of a hot carrageenan solution (D). For solutions, the Newtonian liquid behavior is found at lower frequencies with $G'' > G'$ and moduli increasing with frequencies with the following power laws $G'' \sim \omega$ and $G' \sim \omega^2$. For the dilute solution, the high-frequency regime shows the hydrodynamic behavior with again $G'' > G'$ and a single power law behavior $G' \sim G'' \sim \omega^v$, where v is the exponent describing the solvent–polymer interactions. The rheological signature of polymer entanglements is the crossover between G' and G'' and the occurrence of an elastic plateau at frequencies larger than the Newtonian behavior. The mechanical spectrum of a gel shows solid character with G' independent of frequency and much larger than G''. Note that G'' shows a minimum which is ubiquitous in carrageenan gels and suggests the existence of two relaxation processes at smaller and larger frequencies. In the cooling test pictured in (D), the hot solution is described by $G' < G''$. Then, the gel point is reached at a temperature T_g for which $G' = G''$. Upon further cooling, the gel becomes more elastic as a result of continuous aggregation of helices with time and G' becomes much larger than G''. In the gel state, G' is not frequency dependent. Thus, a temperature sweep performed with another frequency (see thick line in D) should return identical G' values in the gel state, whereas the frequency dependence of the liquid state should be recovered at higher temperatures. The periodic strain signal in the inset of (D) calls the attention for a possible experimental pitfall when the cooling rate is much faster than the time needed (the inverse of the frequency of sinusoidal oscillation) to output G' and G''. Since the reading is no more a sinusoidal, erroneous signal amplitude and phase shift will be measured resulting in untrue G' and G''. This also points to the experimental difficulty in actually determining the gel point, as the transition can be faster than the time needed to measure G' and G''.

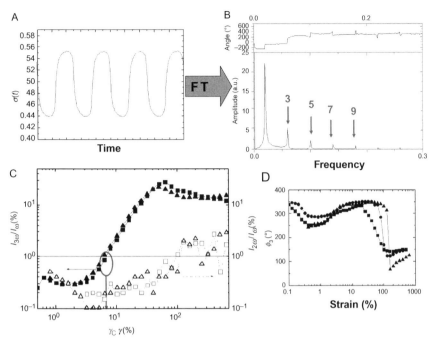

Figure 2.7 Fourier transform rheology (FTR). A nonsinusoidal stress response (A) to a sinusoidal strain in the time space is Fourier transformed to compute the spectra of harmonic amplitudes and phases (B). In the frequency space, harmonics appear at odd frequencies of the applied stress frequency, and the FTR analysis of LAOS relies on computing the scaled intensities of harmonics with respect to the fundamental I_{nw1}/I_{w1} ($n=3, 5, 7, \ldots$), as well as the harmonics phases rescaled with respect to the phase of the fundamental $\phi_{nw1} - n\phi_{w1}$. Then, the nonlinear behavior of the material is described by the strain dependence of scaled harmonics amplitudes (C) and phases (D).

equilibrium gel conditions are reached if enough time is left for the completion of the gel structure. The kinetics of gel formation can actually be followed by performing a time sweep after the cooling: G' and G'' are monitored as a function of time using one stress amplitude (in the linear regime) oscillating at a specific frequency. Details about such experimental protocols can be found elsewhere (see for instance Hilliou & Gonçalves, 2007). An issue occurring during gel formation is water release (syneresis) from the gel. As water is expelled from the gel, a liquid layer will form between the gel and the top disk. In this case, the water plus gel system resembles a Maxwell model. Thus, a liquid response at low frequencies should be measured, and there is no direct access to the gel elastic modulus. Alternatively, one might glue sand paper on the disk to entrap released water into holes, while maintaining disk contact with the gel. This partially solves for the liquid layer problem and slip between gel and disk, but the solution comes with a cost: again, there is a systematic error on the gel elasticity as the actual surface probed to convey the stress is smaller than the disk surface. All these issues, together with variability in carrageenan samples due to the hybrid character, variability in seaweeds, and salt effect (see Sections 4.4 and 4.5), may explain why many different concentration dependences can be found in the literature for the gel elastic modulus of K and I (see for instance Chen, Liao, & Dunstan, 2002; Hossain, Miyanaga, Maeda, & Nemoto, 2001; Plashchina, Grinberg, Braudo, & Tolstoguzov, 1981; Rochas et al., 1989; and references therein).

4.3. Penetration tests

A penetration test consists in plunging a cylinder in a gel and recording the force as a function of the distance traveled in the gel. The distance can be converted in strain, and an apparent Young modulus can be computed from the force reading if some geometric constraints for plunger and gel diameters and gel thickness are respected, and if buoyancy effects are neglected (Oakenfull, Parker, & Tanner, 1989). Stress strain curves measured with penetration test performed in KI and KIMN can be found elsewhere (Azevedo et al., 2014; Hilliou, Larotonda, Abreu, et al., 2006; Hilliou, Larotonda, Sereno, et al., 2006). The test is fast, though slow plunger speed should be chosen for weaker gels, as viscoelasticity will be at play, in contrast to compression tests on solids. The test is easy to implement and many gels can be studied as equilibrated samples can be prepared in advance. However, the test only returns an apparent Young modulus, and the linear regime needed to compute the latter may be difficult to capture, especially for weak gels. In addition, the test gives access to strain hardening, softening, and fracture characteristics. However, water synersis will give same trouble as explained above.

4.4. Effects of salt type and concentration

K is cation specific, whereas I is not (Piculell, 1995). In KCl, K forms stiff and brittle gels, whereas I forms much weaker gels. Thus, the contribution from I blocks to the gel structure and elasticity in KI and KIMN is actually masked by the contributions originating by K blocks. Indeed, the temperature dependence of G' and G'' during the cooling of KIMN and K in KCl shows a single-step increase with cooling and a crossover point indicating the gel

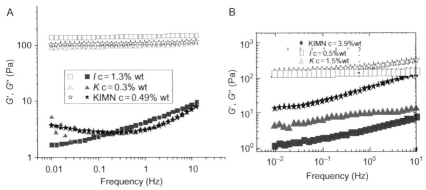

Figure 2.8 (A) Mechanical spectrum of KIMN, K, and I gels formed in 0.05 M KCl using the concentration indicated in the box. (B) Mechanical spectra of the same carrageenan gels but formed in 0.1 M NaCl using the concentration used in the box. Empty symbols show G', and solid symbols show G''. KIMN containing 51 mol% K, 32 mol% I and 17 mol% M and N, were extracted from alkali-treated *M. stellatus* with Na_2CO_3 (Hilliou, Larotonda, Abreu, et al., 2006). K and I are commercial samples from Sigma-Aldrich (Germany).

transition temperature (see for instance Fig. 2.1 in Hilliou, Larotonda, Sereno, et al., 2006). A typical mechanical spectrum of a KIMN gel is displayed in Fig. 2.8A. G' shows a flat plateau, whereas G'' is a decade or so smaller and shows a minimum. K and I gels of matching elasticity are also displayed. They have been obtained using different concentration of K and I only to match the level of gel elasticity. The spectrum of the K gel nicely matches the one of the KIMN, suggesting similar gel structure possibly because of the larger amount in K than in I in the KIMN samples isolated from *M. stellatus* (see caption in Fig. 2.8), whereas the minimum in G'' is shifted to much lower frequency for I. Data in Fig. 2.8A thus suggest that KIMN gels in KCl adopt a structure close to K gels. In NaCl, the situation is changed as a two-step gelation has been reported (Chanvrier et al., 2004; Souza, Hilliou, Bastos, & Gonçalves, 2011). MicroDSC data revealed that the step at higher temperature relates to a coil-to-helix transition assigned to the I blocks, whereas the step increase in G'' at lower temperature relates to the coil-to-helix transition of the K blocks (Souza et al., 2011). In the course of the second step, the KIMN solution gels as indicated by the crossover between G' and G''. The mechanical spectrum of a KIMN gel in NaCl is shown in Fig. 2.8B for a direct qualitative comparison with the KCl case. G' shows a plateau at lower frequency indicating a solid behavior, but at higher frequency, G' grows suggesting a viscoelastic behavior or viscous contributions to the mechanical response. The viscous character is also mirrored in the loss modulus which is an increasing function of the frequency with a power law behavior of 0.3 matching the one of the storage modulus. This is the signature of a weak gel at higher frequency, and not orientational structuring, because in the latter case an exponent close to 0.5 is expected (Larson, 1999). Gels of matching elasticity were also formed with K and I under identical salt conditions, and the mechanical spectra are also reported in Fig. 2.8B. It is clear that the macroscopic structure of the KIMN gel matches the one of K or I (see similar plateau in G' at low frequency). However, the high-frequency regime of KIMN shows differences with respect to K and I counterparts. Note also that the KIMN concentration needed to form a gel with an elasticity of the order of 100 Pa is much higher than the concentrations used for K and I. The chemical structure of KIMN should thus be called to elucidate this. In KCl, K forms a gel with only 60% of the concentration needed to form a KIMN gel. This is to be compared with the 51 mol% of K contained in KIMN. In NaCl, I forms a gel of similar elasticity with 28% of the concentration needed for KIMN to gel. This compares well with the 32 mol% content in the KIMN.

When no salt is used, KIMN can show a gel behavior, but large concentration and low temperature are necessary. This is illustrated in Fig. 2.9 for the same KIMN as in Fig. 2.8, which now is gelled with distilled water. The cooling of the hot solution shows two steps as for NaCl; however, no micro-DSC was run under the same cooling transition to assess coil-to-helix transitions of I and K blocks. However, based on data reported for NaCl, and noting that the KIMN sample possesses roughly 3 wt% Na^+ due to the extraction process with Na_2CO_3 (Souza et al., 2011), one may conjecture that the step at highest temperature corresponds to the formation of I helices from the I blocks, whereas the step increase of moduli at lower temperature relates to the helix formation of K blocks. Figure 2.9 also shows that gel sets only after the second step increase of moduli and thus relates to the aggregation of K helices. This stems from the fact that the phase angle passes through 45° and goes down to 0°, indicating a solid behavior. The inset in Fig. 2.9 shows the mechanical spectrum of the gel formed in water at 5 °C. A soft gel is obtained with a modulus of the order of 2 Pa and a high-frequency regime showing a power law behavior of the order of 0.75. This is indicative of a gel close to the percolation point (Gallani, Hilliou, & Martinoty, 1994).

Gel elasticity of KI depends on ionic strength. Phase diagrams in NaCl and KCl were recently constructed (Azevedo et al., 2014). The diagrams are different from the mixture of K and I with corresponding chemical

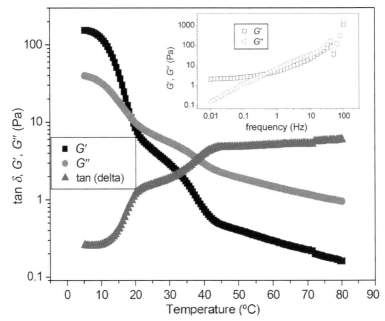

Figure 2.9 Temperature sweep (cooling) performed at 1 Hz and a strain of 5% (linear regime) on a hot distilled water solution of 4 wt% KIMN alkali extracted from *M. stellatus* using Na_2CO_3. Inset: mechanical spectrum recorded under equilibrium (after 24 h) at 5 °C.

composition, that is, 70 wt% K and 30 wt% I, to match the chemical composition of a KI alkali extracted from *M. stellatus* using either NaOH or KOH, in order to not mix salt type during gel formation in NaCl or KCl (Azevedo et al., 2013). The Young modulus of KI gels show a maximum with the ionic strength controlled by NaCl or KCl. The maximum correlates with a structural transition since more turbid gels were obtained at larger ionic strengths. Overall, the study showed that KI gels can accommodate more KCl or NaCl than the corresponding mixture of K and I, since no phase separation was observed nor water syneresis, which can be attractive for specific applications requiring large salt amount.

4.5. Effects of hybrid carrageenan concentration

The gel elastic modulus increases with the KI or KIMN concentration. A power law describes this concentration dependence as indicated by the log–log plots shown in Fig. 2.10 for a KIMN extracted from *M. stellatus* and gelled in KCl. The hybrid carrageenan gel elasticity is compared with

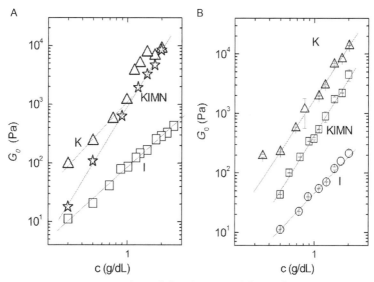

Figure 2.10 Concentration scaling of the elastic moduli G_0 of KIMN gels (stars) from *M. stellatus*, gels from K (triangles), and gels from I (squares) formed in 0.05 M KCl. Lines are power law fits to the data and computed exponents are reported in Table 2.1.

K and I counterpart. Measurements were performed with a dynamic stress rheometer (ARG2, TA Instruments Inc., USA) equipped with parallel plates. Cooling rate used to gel hot solutions was 5 °C/min, and gels were equilibrated with time at 20 °C. For K gels, a serrated plate was used to diminish slip originating from water syneresis. However, for larger concentrations, the gel elastic modulus G_0 measured at 1 Hz departs from the power law behavior and eventually shows a plateau, which might be related with slip problems or structural transition. Overall, KIMN show G_0 values between K and I G_0 values, which is consistent with the industrial denomination of "weak kappa" or "kappa-2." Power law exponents are indicated in Fig. 2.10A and are also gathered in Table 2.1, to offer a direct comparison with exponents computed from the concentration dependence of G_0 measured with the same set of carrageenans but using a much slower cooling rate, that is, 1 °C/min. Corresponding data are plotted in Fig. 2.10B. Cooling rate does not affect the exponent computed for KIMN in contrast to both I and K and as reported in the literature (Piculell, 1995). However, more elastic KIMN and K gels are obtained with slower cooling, which agrees with earlier reports for K (see for instance Piculell, 1995). Note that for large concentrations, K gels cooled at the slowest rate show a power law behavior. Experiments were conducted with a hatched plate use to avoid slip of polymer melts, in contrast to the serrated plate used for data in Fig. 2.10A. This different experimental setup might explain the absence of plateau. Alternatively, the cooling rate and the effect of thermal history could explain the difference originating from different gel structure. Table 2.1 also gives values of exponents obtained from penetration tests carried out on gels formed KCl with KI extracted from *M. stellatus* using a harder alkali treatment (Azevedo et al., 2014). Exponents are compared with those computed from data obtained with mixtures of K and I using the same K and I proportions matching the chemical structure of KI. Different exponents are found for KI and mixtures of K and I again pointing to the fact that the chemical structure of KI is that of a block copolymer.

4.6. Relationships between the chemical structure and the gel properties

Since the phase diagram of a KI shows that gel properties are dependent on KI concentration and salt type and concentration, one needs to assess the effect of KI chemical composition on gel elasticity with caution. This was carried out by van de Velde et al. (2005) for a set of KI free of impurities (such as pyruvate and Floridean starch) and gelled in identical conditions, that is, in the presence of same salt mixture (13.4 mM K^+ and 6.8 mM Ca^{2+}) and using 1 wt% hybrid carrageenan. An increase of gel elasticity with the relative content in K in the hybrid carrageenan was found, and a nearly linear relationship was established for K contents above 50 mol%. Below this critical content, weak gels were found despite the large amounts of I. This is of course due to the choice of the salt mixture used to essentially assess the gelling effect of K (Parker et al., 1993). Thus, the study is giving only a partial picture of the effect of KI chemical composition on gel elasticity. A quite different conclusion was reached with KIMN extracted from a single seaweed using different mild Na_2CO_3 treatments. The storage modulus G_0 of gels formed with 1.5 wt% KIMN in 0.05 M KCl at 20 °C (Hilliou, Larotonda, Abreu, et al., 2006; Hilliou, Larotonda, Sereno, et al., 2006) does not correlate with the relative content in K, M, or N measured with ^{1}H NMR. A negative correlation was found between G_0 and the sulfate degree of KIMN measured with FTIR, which is in agreement with similar correlation found with agar gels (Lahaye, 2001). Interestingly, the molecular mass of KIMN was found to have an impact on G_0 as longer chains form gels with smaller G_0. It was concluded that a complex interplay between chemical composition and structure of KIMN and the gel properties was at play, with various compensating effect. Thus, another study was conducted on selected KIMN where either the relative content in I or the molecular mass was varied while keeping all other KIMN characteristics constant. Also, gel properties were studied under different gel conditions, that is, 2 wt% hybrid carrageenan in deionized water, 0.1 and 1 M NaCl solutions. It was concluded that the molecular mass affects the gel elasticity of KIMN and that it is not enough to solely consider the relative content in gelling carrageenan in the hybrid to unravel the gel elastic properties. More recently, KI and KIMN obtained from the same seaweeds but using a stronger alkali treatment with NaOH or KOH were tested for their gel elasticity (Azevedo et al., 2013). Results pointed to the fact that KIMN form weaker gels than KI, either in NaCl or in KCl. KI with similar chemical composition (68 mol%K and 32 mol% I) but decreasing length show gels with weaker elasticity in contrast to results obtained with KIMN. This suggests that more precise information on the block length is needed to explain the results, but this copolymer characteristic is hard to assess either enzymatically (since this only allows assessing the chemical composition of degradation products, i.e., oligomers) or with spectroscopic chemical techniques. (These only give information on short lengths of few chemical bonds.)

4.7. Large deformation behavior and gels under steady flow

In a seminal review paper on KI, van de Velde (2008) underlined the lack of characterization of KI gels in large deformation and lack of data in the literature for instance for strain at break. Since then, studies were carried out in our group to screen for the nonlinear properties of KIMN (Hilliou et al., 2009). LAOS studies using FTR showed that KIMN gels are strain hardening and that the strain hardening shows scaling behavior with the KIMN concentration. Since such scaling also exists in colloidal gels, the rationalization of FTR data by models devised for colloidal systems with aggregation driven by diffusion or chemical reaction. This allowed for the computation of the fractal dimension of the gels which was shown to match the fractal dimension obtained with cryoSEM imaging. The set of results showed that KIMN gels are essentially made of rigid rodlike fibrils which connect to build a gel with a fractal dimension of the order of 1.67. Strain hardening comes from the fibril structure and is ubiquitous in biological gels (Dobrynin & Carrillo, 2011). Penetration tests performed on a set of KI gels formed at different concentrations and under different salt conditions confirmed the strain-hardening character (Azevedo et al., 2014). Finally, the effect of steady shear applied during the cooling of a gel forming of a KIMN solution has been recently studied (Hilliou, Sereno, & Gonçalves, 2014). The gels formed under such conditions show a G_0 which is a decreasing function of the applied shear rate. In addition, the steady shear did not affect the temperature at which helix formation is initiated. Polarized optical microscopy revealed the orientation of structure in these gels. Based on this experimental evidence, a new model for the gel formation under steady shear has been proposed.

5. PERSPECTIVES

All the seaweeds available for KI and KIMN production are obtained from the harvesting of wild species. A sustainable production of the natural resource requires adopting the same cultivation strategies as for K and I. Actually, *C. crispus* is now farmed in Canada, and successful integrated multitrophic aquaculture of *C. crispus* has recently been demonstrated (Abreu, Pereira, Yarish, Buschmann, & Sousa-Pinto, 2011; Domingues, 2010). However, more Gigartinales must be tested to extend the palette of hybridicity in KI. The interplay between the chemical structure of gelling hybrid carrageenan and the gel elastic properties is not yet unraveled, as too few reports with contradictory conclusions are available in the literature. More effort is needed for the careful gel characterization under suited carrageenan concentration and salt conditions as salt may express more the K characteristic or the I characteristic of the KI and KIMN, thus modifying the overall picture. Finally, gel behavior under large and fast deformations is still largely unexplored. There is a need for reports in this area as new products could emerge from the combined investigation of KI with different chemistry and of nonlinear rheological behavior.

ACKNOWLEDGMENTS
This work was supported by FEDER funds through the program COMPETE (project PTDC/CTM/100076/2008) and by the Portuguese Foundation for Science and Technology (PEst-C/CTM/LA0025/2013—Projecto Estratégico—LA 25—2013–2014—Strategic Project—LA 25—2013–2014).

REFERENCES
Abreu, M. H., Pereira, R., Yarish, C., Buschmann, A. H., & Sousa-Pinto, I. (2011). IMTA with Gracilaria vermiculophylla: Productivity and nutrient removal performance of the seaweed in a land-based pilot scale system. *Aquaculture*, *312*, 77–87.
Amimi, A., Mouradi, A., Bennasser, L., & Givernaud, T. (2007). Seasonal variations in thalli and carrageenan composition of *Gigartina pistillata* (Gmelin) Stackhouse (Rhodophyta, Gigartinales) harvested along the Atlantic coast of Morocco. *Phycological Research*, *55*, 143–149.

Azevedo, G., Bernardo, G., & Hilliou, L. (2014). NaCl and KCl phase diagrams of kappa/iota-hybrid carrageenans extracted from *Mastocarpus stellatus*. *Food Hydrocolloids, 37*, 116–123.

Azevedo, G., Hilliou, L., Bernardo, G., Sousa-Pinto, I., Adams, R. W., Nilsson, M., et al. (2013). Tailoring kappa/iota-hybrid carrageenan from *Mastocarpus stellatus* with desired gel quality through pre-extraction alkali treatment. *Food Hydrocolloids, 31*, 94–102.

Barnes, H. A., & Bell, D. (2003). Controlled-stress rotational rheometry: An historical review. *Korea–Australia Rheology Journal, 15*, 187–196.

Barnes, H. A., Hutton, J. F., & Walters, K. (1989). *An introduction to rheology*. Amsterdam, The Netherlands: Elsevier Science Publishers.

Bixler, H. J. (1996). Recent developments in manufacturing and marketing carrageenan. *Hydrobiologia, 326*(327), 35–57.

Bixler, H. J., Johndro, K., & Falshaw, R. (2001). Kappa-2 carrageenan: Structure and performance of commercial extracts II. Performance in two simulated dairy applications. *Food Hydrocolloids, 15*, 619–630.

Bixler, H. J., & Porse, H. (2011). A decade of change in the seaweed hydrocolloids industry. *Journal of Applied Phycology, 23*, 321–335.

Chanvrier, H., Durand, S., Garnier, C., Sworn, G., Bourriot, S., & Doublier, J.-L. (2004). Gelation behaviour and rheological properties of k/i-hybrid carrageenans. In P. A. Williams, & G. O. Phillips (Eds.), *Gums and stabilisers for the food industry: Vol. 12.* (pp. 139–144). Cambridge: The Royal Society of Chemistry.

Chen, Y., Liao, M.-L., & Dunstan, D. E. (2002). The rheology of K^+-kappa-carrageenan as a weak gel. *Carbohydrate Polymers, 50*, 109–116.

Chopin, T., Kerin, B. F., & Mazerolle, R. (1999). Phycocolloid chemistry as a taxonomic indicator of phylogeny in the Gigartinales, Rhodophyceae: A review and current developments using Fourier transform infrared diffuse reflectance spectroscopy. *Phycological Research, 47*, 167–188.

Dobrynin, A. V., & Carrillo, J.-M. Y. (2011). Universality in nonlinear elasticity of biological and polymeric networks and gels. *Macromolecules, 44*, 140–146.

Domingues, B. (2010). *Bioremediation efficiency of Mastocarpus stellatus (Stackhouse) Guiry in an integrated multi-trophic aquaculture*. Master Thesis, Spain: University of Barcelona.

Fogg, G. E. (1964). Environmental conditions and the pattern of metabolism in algae. In D. Jackson (Ed.), *Algae and man* (pp. 77–85). New York: Plenum Press.

Freile-Pelegrin, Y., & Robledo, D. (2006). Carrageenan of *Eucheuma isiforme* (Solieriaceae, Rhodophyta) from Yucatán, Mexico. II. Seasonal variations in carrageenan and biochemical characteristics. *Botanica Marina, 49*, 72–78.

Fuller, S. W., & Mathieson, A. C. (1972). Ecological studies of economic red algae. IV. Variations of carrageenan concentration and properties in Chondrus crispus Stackhouse. *Journal of Experimental Marine Biology and Ecology, 10*, 49–58.

Gallani, J.-L., Hilliou, L., & Martinoty, P. (1994). Abnormal viscoelastic behavior of side-chain liquid-crystal polymers. *Physical Review Letters, 72*, 2109–2112.

Garrec, D. A., & Norton, I. T. (2012). Understanding fluid gel formation and properties. *Journal of Food Engineering, 112*, 175–182.

Guibet, M., Boulenguer, P., Mazoyer, J., Kervarec, N., Antonopoulos, A., Lafosse, M., et al. (2008). Composition and distribution of carrabiose moieties in hybrid κ-/ι-carrageenans using carrageenases. *Biomacromolecules, 9*, 408–415.

Hilliou, L., & Gonçalves, M. P. (2007). Gelling properties of a kappa/iota-hybrid carrageenan: Effect of concentration and steady shear. *International Journal of Food Science and Technology, 42*, 678–685.

Hilliou, L., Larotonda, F. D. S., Abreu, P., Abreu, M. H., Sereno, A. M., & Gonçalves, M. P. (2012). The impact of seaweed life phase and postharvest storage duration on the chemical and rheological properties of hybrid carrageenans isolated from Portuguese *Mastocarpus stellatus*. *Carbohydrate Polymers, 87*, 2655–2663.

Hilliou, L., Larotonda, F. D. S., Abreu, P., Ramos, A. M., Sereno, A. M., & Gonçalves, M. P. (2006). Effect of extraction parameters on the chemical structure and gel properties of kappa/iota-hybrid carrageenans obtained from *Mastocarpus stellatus*. *Biomolecular Engineering, 23*, 201–208.

Hilliou, L., Larotonda, F. D. S., Sereno, A. M., & Gonçalves, M. P. (2006). Thermal and viscoelastic properties of kappa/iota-hybrid carrageenan gels obtained from the Portuguese seaweed *Mastocarpus stellatus*. *Journal of Agricultural and Food Chemistry, 54*, 7870–7878.

Hilliou, L., Sereno, A. M., & Gonçalves, M. P. (2014). Gel setting of hybrid carrageenan solutions under steady shear. *Food Hydrocolloids, 35*, 531–538.

Hilliou, L., Wilhelm, M., Yamanoi, M., & Gonçalves, M. P. (2009). Structural and mechanical characterization of kappa/iota-hybrid carrageenan gels in potassium salt using Fourier transform rheology. *Food Hydrocolloids, 23*, 2322–2330.

Hossain, K. S., Miyanaga, K., Maeda, H., & Nemoto, N. (2001). Sol-gel transition behavior of pure iota-carrageenan in both salt-free and added salt states. *Biomacromolecules, 2*, 442–449.

Hyun, K., Wilhelm, M., Klein, C. O., Cho, K. S., Nam, J. G., Ahn, K. H., et al. (2011). A review on nonlinear oscillatory shear tests: Analysis and application of large amplitude oscillatory shear (LAOS). *Progress in Polymer Science, 36*, 1697–1753.

Lahaye, M. (2001). Developments on gelling algal galactans, their structure and physicochemistry. *Journal of Applied Phycology, 13*, 173–184.

Larson, R. G. (1999). *The structure and rheology of complex fluids*. New York: Oxford University Press.

McCandless, E. L. (1981). Polysaccharides of seaweeds. In C. S. Lobban, & M. J. Wynne (Eds.), *The biology of seaweeds* (pp. 559–588). Oxford: Blackwell Scientific Publications.

McHugh, D. J. (2003). *A guide to the seaweed industry*. Rome: FAO Fisheries, Technical Paper No. 441.

Mendoza, W. G., Ganzon-Fortes, E. T., Villanueva, R. D., Romero, J. B., & Montaño, M. N. E. (2006). Tissue age as a factor affecting carrageenan quantity and quality in farmed *Kappaphycus striatum* (Schmitz) Doty ex Silva. *Botanica Marina, 49*, 57–64.

Oakenfull, D. G., Parker, N. S., & Tanner, R. I. (1989). Method for determining absolute shear modulus of gels from compression tests. *Journal of Texture Studies, 19*, 407–417.

Parker, A., Brigand, G., Miniou, C., Trespoey, A., & Vallée, P. (1993). Rheology and fracture of mixed i- and k-carrageenan gels: Two-step gelation. *Carbohydrate Polymers, 20*, 253–262.

Pereira, L., & Mesquita, J. F. (2003). Carrageenophytes of occidental Portuguese coast. 1-Spectroscopic analysis in eight carrageenophytes from Buarcos bay. *Biomolecular Engineering, 20*, 217–222.

Pereira, L., & Mesquita, J. F. (2004). Population studies and carrageenan properties of *Chondracanthus teedei* var. *lusitanicus* (Gigartinaceae, Rhodophyta). *Journal of Applied Phycology, 16*, 369–383.

Piculell, L. (1995). Gelling carrageenans. In A. M. Stephen (Ed.), *Food polysaccharides and their applications* (pp. 205–244). New York: Marcel Dekker.

Plashchina, I. G., Grinberg, N. V., Braudo, E. E., & Tolstoguzov, V. B. (1981). Viscoelastic properties of kappa-carrageenan gels. *Colloid & Polymer Science, 258*, 939–943.

Rees, D. A. (1972). Shapely polysaccharides. *Biochemical Journal, 126*, 257–273.

Rees, D. A., Morris, E. R., & Robinson, G. R. (1980). Cation specific aggregation of carrageenan helices: Domain model of polymer gel structure. *Journal of Molecular Biology, 138*, 349–362.

Rochas, C., Rinaudo, M., & Landry, S. (1989). Relation between the molecular structure and gel mechanical properties of carrageenan gels. *Carbohydrate Polymers, 10*, 115–127.

Romero, J. B., Villanueva, R. D., & Montano, M. N. E. (2008). Stability of agar in the seaweed *Gracilaria eucheumatoides* (Gracilariales, Rhodophyta) during postharvest storage. *Bioresource Technology, 99*, 8151–8155.

Souza, K. K. S., Hilliou, L., Bastos, M., & Gonçalves, M. P. (2011). Effect of molecular weight and chemical structure on thermal and rheological properties of gelling kappa/iota-hybrid carrageenan solutions. *Carbohydrate Polymers, 85*, 429–438.

Stancioff, D. J. (1981). Reflections on the interrelationships between red seaweed source, chemistry and use. *Proceedings of the International Seaweed Symposium: 10.* (pp. 113–121).

van de Velde, F. (2008). Structure and function of hybrid carrageenans. *Food Hydrocolloids, 22*, 727–734.

van de Velde, F., Antipova, A. S., Rollema, H. S., Burova, T. V., Grinberg, N. V., Pereira, L., et al. (2005). The structure of kappa/iota-hybrid carrageenans. II. Coil-helix transition as a function of chain composition. *Carbohydrate Research, 340*, 113–1129.

van de Velde, F., Peppelman, H. A., Rollema, H. S., & Tromp, R. H. (2001). On the structure of kappa/iota-hybrid carrageenans. *Carbohydrate Research, 331*, 271–283.

Viebke, C., Piculell, L., & Nilsson, S. (1994). On the mechanism of gelation of helix-forming biopolymers. *Macromolecules, 27*, 4160–4166.

Villanueva, R. D., Hilliou, L., & Sousa-Pinto, I. (2009). Postharvest culture in the dark: An eco-friendly alternative to alkali treatment for enhancing the gel quality of kappa/iota-hybrid carrageenan from *Chondrus crispus* (Gigartinales, Rhodophyta). *Bioresource Technology, 100*, 2633–2638.

Villanueva, R. D., Mendoza, W. G., Rodrigueza, M. R. C., Romero, J. B., & Montaño, M. N. E. (2004). Structure and functional performance of gigartinacean kappa–iota hybrid carrageenan and solieriacean kappa–iota carrageenan blends. *Food Hydrocolloids, 18*, 283–292.

Wakibia, J. G., Bolton, J. J., Keats, D. W., & Raitt, L. M. (2006). Seasonal changes in carrageenan yield and gel properties in three commercial Eucheumoids grown in southern Kenya. *Botanica Marina, 49*, 208–215.

CHAPTER THREE

Isolation of Low-Molecular-Weight Heparin/Heparan Sulfate from Marine Sources

Ramachandran Saravanan[1]
Department of Marine Pharmacology, Faculty of Allied Health Sciences, Chettinad Hospital and Research Institute, Chennai, India
[1]Corresponding author: e-mail address: saran_prp@yahoo.com

Contents

1. Introduction	46
2. History of Heparin	47
3. Anticoagulant Activity of Heparin	48
4. Sources of Heparin	49
4.1 Terrestrial sources	49
4.2 Marine sources	50
5. Difference Between Heparin and HS	52
6. Biomedical Significance of LMWH/HS	53
7. Isolation of LMWH Sulfate	54
7.1 Chemical synthesis of low molecular heparin/HS	54
7.2 Enzymatic synthesis of low molecular heparin/HSs	55
7.3 Chromatography separation of LMWH/HS	56
8. Conclusion	58
References	59

Abstract

The glycosaminoglycan (heparin and heparan sulfate) are polyanionic sulfated polysaccharides mostly recognized for its anticoagulant activity. In many countries, low-molecular-weight heparins have replaced the unfractionated heparin, owing to its high bioavailability, half-life, and less adverse effect. The low-molecular-weight heparins differ in mode of preparation (chemical or enzymatic synthesis and chromatography fractionations) and as a consequence in molecular weight distribution, chemical structure, and pharmacological activities. Bovine and porcine body parts are at present used for manufacturing of commercial heparins, and the appearance of mad cow disease and Creutzfeldt–Jakob disease in humans has limited the use of bovine heparin. Consequently, marine organisms come across the new resource for the production of low-molecular-weight heparin and heparan sulfate. The importance of this chapter suggests that the low-molecular-weight heparin and heparan sulfate from marine species could be alternative sources for commercial heparin.

1. INTRODUCTION

Heparin is a polydisperse mixture of linear glycosaminoglycan (GAG) termed as heteropolysaccharides, a component that is comprised of highly sulfated (1 → 4) linked uronic acid–(1 → 4)-D-glucosamine repeating disaccharide units, secreted by the mast cells of mammals, and is found in the tissues of lymph nodes, skin, intestines, and lungs (Guo et al., 2003; Saravanan & Shanmugam, 2011). Heparin is a highly negatively charged GAG whose structure is a highly complex structure (Fig. 3.1), disaccharide units from 10 to 50 of which encompass most heparin samples (Monique, Kher, & Toulemonde, 1999). The uronic acid usually comprises 90% L-idopyranosyluronic acid (L-iduronic acid) and 10% D-glucopyranosyluronic acid (D-glucuronic acid). At the disaccharide level, a number of structural variations exist, leading to sequence microheterogeneity within heparin, and have a molecular mass range of 50–150 kDa, an average molecular mass of ~75 kDa.

Heparin's complexity extends through multiple structural intensities. At the proteoglycan (PG) level, different numbers of GAG chains (possibly having different disaccharide sequences) can be attached to the various serine residues present on the core protein. Heparin chains are biosynthesized attached to a unique nucleus protein, serglycin, found primarily in mast cells. On mast cell degranulation, proteases act on the heparin core protein to release peptidoglycan heparin, which is further processed by a β-endoglucuronidase into GAG heparin (Linhardt & Toida, 2004). The chemical, physical, and biological properties of heparin are primarily attributed to GAG sequence, disaccharide conformation, chain flexibility, molecular weight, and charge density. Heparan sulfate (HS), though structurally associated to heparin, is a very smaller amount surrogated accompanied by sulfate clusters and has a more varied structure. D-Glucuronic acid

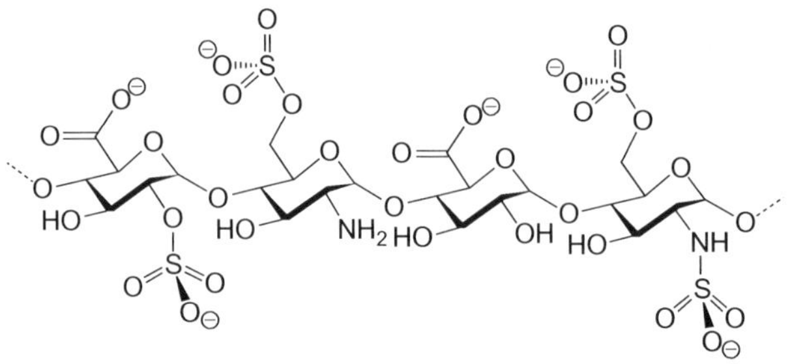

Figure 3.1 Heparin structure.

preponderates in HS, and it is also polydisperse, having a low molecular mass of ~30 kDa ranging from 5 to 50 kDa. HS chains also habitually contain domains of extended chains having low or high sulfation (Gallagher, Turnbull, & Lyon, 1992).

Since heparin was discovered more than 50 years ago, our knowledge in the field of chemical structure and molecular interactions of this attracting polymer is uncompleted at the early phases of its development. However the attempts of a most important multidisciplinary cluster of investigators and clinicians, it is now well distinguished that heparin has multiple sites of reactions and can be used in quite a lot of gestures. It is not too distant in the future to observe the impact of heparin on the organization of various ailments (Linhardt & Gunay, 1999). It has been used for over half a century as an anticoagulant and antithrombotic drug. Approximately 500 million heparin doses are currently used worldwide in each year for this purpose, and the large number of favorable biological activities unrelated to anticoagulant leads to a constantly growing interest for new therapeutic applications (Linhardt, 1991; Lindahl, Lidholt, Spillmann, & Kjellen, 1994).

2. HISTORY OF HEPARIN

The heparin is a clinical anticoagulant and has been one of the most successful and widely used drugs of this century (Linhardt & Toida, 1997). It is routinely prescribed before most major surgeries, which was discovered by Mc Lean (1916). Earlier, it was termed as natural anticoagulant, and later named as heparin. McLean determined the anticoagulant activity of heparin for several times and he became satisfied that heparin did actually inhibit the coagulation of the serum–plasma test mixture as well as whole blood *in vitro*; later, he determined the anticoagulant activity of heparin by administering the heparin intravenously to dog (first *in vivo* experiment). Awareness of the chemical structure and molecular interactions of this captivating polycomponent was partial at the early stages of its progress. Through the efforts of a key multidisciplinary group of researchers and clinicians, it is now well recognized that heparin has multiple sites of actions and can be used as a marker for various indications. The first intravenous clinical usage of heparin product is saturated from 1936 (Best, 1959).

Heparin is not absorbed from the gastrointestinal tract and must be administered parenteral, and heparin has no oral bioavailability, presumably because of its size and ionic repulsion from negatively charged epithelial tissue. The onset action of heparin is immediate by administrating heparin intravenously. However, the absorption usually occurs within 20–60 min of administration through subcutaneous injection of heparin (Takats, 1950). PGs originated in the membranes of every animal cell, intracellularly in secretory granules of preferred cells or extracellular in the milieu, where they exhibit a broad range of biological functions. Even though the core protein of PGs is important, various actions interceded by PGs are supposed to result from their GAG linear chains. Among the finding of the anticoagulant activity of heparin, attention in the other associates of the GAG family unit has improved; as a result, HS, hyaluronic acid (HA), chondroitin sulfate (CS), and dermatan sulfate (DS) became a spotlight of several examinations (Karst & Linhardt, 2003). The difficulty of GAG polymers creates the clarification of their complicated structure, and the structural sources for many of their imperative biological actions are silent vestiges vague. Earlier, Linhardt and Gunay (1999) have proved that heparin is made up of major component of an uronic acid and O-sulfated glucosamine residues as a component and generalized structure of heparin is drawn in 1970.

3. ANTICOAGULANT ACTIVITY OF HEPARIN

The activity of heparin can be directed by coagulation times and it is widely used as a clinical anticoagulant for the treatment and prevention of thrombotic diseases and for maintaining blood fluidity in extracorporeal devices. By potentiating the activity of serine protease inhibitors such as antithrombin III (AT III) and heparin cofactor II, heparin inhibits serine proteases and blocks the coagulation cascade in blood. Heparin also acts as an antithrombotic agent through its action on platelets and on the endothelium of various tissues, as measured by its *in vivo* blocking of thrombus formation. The anticoagulant action of pharmaceutical heparin (120–180 USP U/mg) is the most thoroughly studied report of its activities. Anticoagulation occurs when heparin binds to AT III, a serine protease inhibitor (serpin). AT III undergoes a conformational change and becomes activated as an inhibitor of thrombin and other serine proteases in the coagulation cascade (Gordon & Strickland, 1989).

From Fig. 3.2, the heparin boosts the rate at which AT III neutralizes thrombin and activated factor X (Xa). More than 60% of the polysaccharides have molecular masses between 2 and 8 kDa, resulting in a reduction in thrombin-neutralizing capacity (i.e., anti-IIa activity). AT III also neutralizes other activated coagulation factors, like factors IX, XI, XII, and plasmin. With low-dose heparin therapy, anticoagulation appears to result from

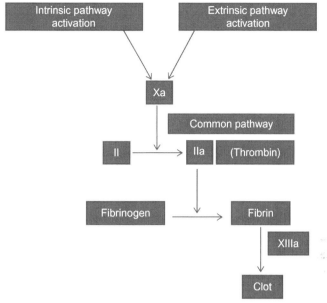

Figure 3.2 Intrinsic/extrinsic pathways of the coagulation cascade leading to fibrin generations.

neutralization of Xa, which converts the alteration of prothrombin to thrombin. The anti-IIa-to-anti-Xa ratio for unfractionated heparin (UFH) is roughly one. By convention, the anti-Xa activity of low-molecular-weight heparins (LMWHs) is indistinguishable, although the anti-IIa decreases as the heparin molecular weight decreases. With full-dose heparin therapy, anticoagulation appears to result primarily from neutralization of thrombin which prevents the conversion of fibrinogen to fibrin (Gordon & Strickland, 1989).

4. SOURCES OF HEPARIN

4.1. Terrestrial sources

Recent statistics use of heparin shows that there are about approximately 10–12 million Americans who are treated with heparin every year. The annual worldwide consumption of crude heparin is over 200 tons and vending revenues are in excess of $5.0 billion per year (Cox, 2004; Bhattacharya, 2008; Wassener, 2010). Currently, bovine/porcine lung, kidney, liver, spleen, pancrease, and intestines are used as raw materials for preparation of commercial and pharmaceutical heparins (Linhardt & Gunay, 1999).

These tissues are rich in mast cells, presumably resulting from the high foreign parasite burden in these tissues, for example, bacteria and viruses. The manifestation of bovine spongiform encephalopathy, "mad cow disease," and its apparent link to the similar prion-based Creutzfeldt–Jakob disease in humans (Schonberger, 1998) has inadequate use of bovine heparin. Furthermore, it is not trouble-free to make distinction between bovine and porcine heparins, making it difficult to ensure the species source of heparin (Linhardt & Gunay, 1999; Saravanan, Vairamani, & Shanmugam, 2010; Saravanan & Shanmugam, 2010).

Porcine heparin also has problems with its use, associated with religious restrictions among members of the Muslim and Jewish faiths. Heparin exhibits anticoagulant activity primarily from its binding to the serine protease inhibitor, antithrombin (Warda, Gouda, Toida, Chi, & Linhardt, 2003). Nonanimal sources of heparin, such as chemically synthesized, enzymatically synthesized, and recombinant heparins, are currently not available for pharmaceutical purposes. In these concerns, researchers have motivated us to look for alternative, nonmammalian sources of heparin (Saravanan et al., 2010).

4.2. Marine sources

Marine organisms are remaining a largely unexploited reservoir, many of which contain a biotechnology/pharmaceutical application. Many organisms are composed of molecules and materials exhibiting interesting distinctiveness and properties, which comprise an exciting store for the development of novel, medical-orientated value-added products. Similar to their terrestrial sources, marine organisms produce a substantial diversity of biomolecules; within polysaccharides, a particular group—GAG—will be addressed in a different section, once these typically sulfated polymers, with a typically bearings repeating unit constituted by a hexose and a hexosamine, are synthesized in the organism in association with proteins forming the PGs (Ori, Wilkinson, & Fernig, 2011). The sulfated polysaccharides present in marine mollusk are high when compared to bovine mucosal heparin (73.5%) and porcine mucosal heparin (72.8%) (Vidhyanandhini, Saravanan, Vairamani, & Shanmugam, 2014). Polysaccharides are polymers composed of carbohydrate monomers (normally hexoses) linked by glycosidic bonds. All the polysaccharides have similar chemical structures, but the apparently small differences are responsible for distinct properties of the polymers.

A various number of heparin/HS analogs derived from enormous marine organisms have been illustrated (Fig. 3.3). Most of these bioactive compounds have been widely reported in expressions of structure, biochemical and molecular effect, and mechanism of actions on cell receptors and estimated in preclinical examinations in rodent animals with promising outcomes. One of the most important aspects of heparin extracted from marine sources is the low-level risk of contamination with microorganisms like virus and bacteria, as marine organisms are evolutionarily secluded from terrestrial mammals. As a result, the likelihood of microorganism or prion contamination in mammalian cells is very unlikely. An alternative pertinent point concerning the remedial exercise of an animal-extracted medicine is the scientific and financially viable prospect to attain incredibly outsized quantities in an invariable and economically acceptable method. In general, the heparin/HS analogs separated from marine lower invertebrate animals provide a reasonable yield (there are about 0.5–2% of the dry weight, comparing to 0.022% from pig intestinal mucosa), by methods analogous to those already existing and employed in the preparation of pharmaceutical heparin. Several species of marine mollusks (especially bivalves, gastropods, and cephalopods species) and sea cucumbers, including those containing

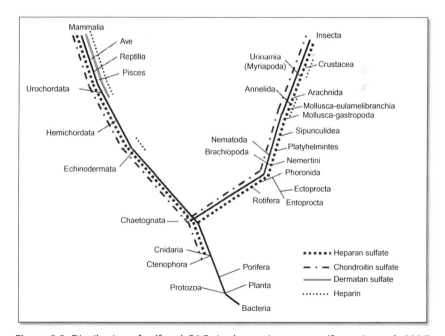

Figure 3.3 Distribution of sulfated GAGs in the marine sources (Sampaio et al., 2006).

high quantities of heparin analogs, have been effectively refined for a long time in different parts of the world. The present aquaculture technologies are capable to produce ton quantities of raw material for the production of heparin from marine sources. The world's production of sea cucumber reached about 21,000 tons in 2001, and that of marine scallops, there are about 73,000 tons in 1999. Possibly, the most important limitations for remedial purpose of polysaccharides from marine organisms are a more profound examination of their properties on mammalian systems and their mechanisms of action compared with heparin and few of them are in clinical trial stages 3–4 (Pavao & Mourao, 2012).

5. DIFFERENCE BETWEEN HEPARIN AND HS

HS is also associated with the GAG family of different chain lengths, molecular weight distribution, and different physiochemical chemical characteristics that result from their various methods of preparation, which create them noninterchangeable. The chemical structure of heparin/HS starts from the primarily produced monotonous polymer being extensively personalized (epimerization and sulfation at different arrangements in the sugar rings). These alterations are substoichiometric and grouped to generate characteristic domains, which vary in size and number within each chain (Stringer & Gallagher, 1997). Nevertheless, HS explains great sequences of (D-glucuronic acid (GlcA) or L-iduronic acid (IdoA)) and D-glucosamine (GlcN) residues, which may be N-sulfated or N-acetylated and which do not exist in heparins [IdoA(2S)–GlcNS(6S)]. In addition, HS also illustrates the lesser degree of sulfation in iduronic acid residue, which is important for the protein binding to cell receptors for most of these biological interactions.

However, the criteria for differencing between HS and heparin are still dubious. The tenure HS has been conventionally used to describe heparinoids (heparin-like byproducts) of the pharmaceutical preparation of heparin from animal sources such as bovine lung or pig mucosa. The HS in these heparinoids had slight or no anticoagulant activity and showed a substantial heterogeneity in molecular weight and sulfate content with the degree of polymer sulfation, although it is currently clear that HS has a distinct structure of heparin and a broader range of biochemical/pharmacological roles in medicine (Gallagher & Walker, 1985; Toida et al., 1997; Linhardt & Claude, 2003).

Most of the structural studies reported in this chapter are based on the specificities of the enzymes prepared from *Flavobacterium heparinum*. Briefly, heparin and HS enzymes are classified as heparinases I, II, and III. The heparinases act upon glucosaminido-iduronic acid linkages where the glucosamine is sulfated at the 2 position. The heparinase I is specific for N-acetyl or N-sulfate glucosaminido-glucuronic acid linkages. This enzyme acts on heparin and HS regions where the N-acetyl or N-sulfate glucosamine is not sulfated at the 6 position. Whereas the heparinase II acts only on heparin, but heparinase III is relatively nonspecific enzyme acting preferentially upon glucosaminido-glucuronic acid linkages where the N-acetyl or N-sulfate glucosamine is sulfated at the 6 position, these enzymes hydrolyze only HS. None of the enzymes act on N-desulfated glucosaminido-iduronic acid linkages (Saravanan, 2010).

6. BIOMEDICAL SIGNIFICANCE OF LMWH/HS

The difference in molecular components and pharmacological functions of LMWHs is replicated in clinical trials that have expressed differences in clinical efficacy and safety. Thus, each LMWH should be believed as an exclusive substance. The important differences between LMWHs are distinctions in the sum of pharmacologically active product, unreliable chemical, and physical properties ensuing in diverse biologic roles, and differences in results from clinical trials performed at optimized doses for each agent. The biological significance of LMWH/HS is used through their competence to engage protein ligand. The results of protein and ligand interactions array from elaborating large-scale structures to controlling the gradient formation and signaling activities of growth factors, cytokines, and morphogens and the localization and activity of many extracellular enzymes (Vyas et al., 2008; Yu et al., 2009). Heparin/HS interaction with protein involves the blood coagulations, hemostasis, fibrinolysis, inflammation against antigens, angiogenesis, defense mechanisms, and endocytosis and inhibits the apoptosis against cancer cells, cell matrix assembly, cell recognition during cell division, cell proliferation, cell invasion, cell migration, and cell adhesion (Sampaio et al., 2006).

LMWH stands for the most considerable development in antithrombotic remedy. Due to the 85–95% of bioavailability after subcutaneous administration, long plasma half-life consequential in an unsurprising response, and less bleeding for a clinically significant antithrombotic activity as compared to UFH, LMWHs are used as potential drugs for antithrombotic agents. The use of LMWH in obstetric patients is owing to the properties of LMWH not crossed the placental blood–barrier and is associated with a better compliance, lesser risk for thrombocytopenia and bleeding complications compared to UFH. Different types of techniques are currently used for the preparation of LMWH from UFH, have variable molecular weight allocation, and consequently are likely to have various pharmacokinetic and pharmacodynamic properties, which may have significant clinical insinuations (Mousa, 2007).

Heparin has great attention and important role in pharmacology including antitumor and antimetastatic activity, and high risk of bleeding complications due to anticoagulant and antithrombotic properties of heparin is very limited usage in cancer treatment. Various research has been done in the previous decades to facilitate imitate and exclusively adapt the biological action of heparin by exogenous heparin species (i.e., UFH, LMWH, heparin oligosaccharides, chemically modified heparins, and biotechnologically obtained heparins). While it will be established, molecular mass, concentration, and charge distribution of heparin play an important role in its biochemical/pharmacological activity. Nowadays, medical information and studies have revealed both the potential and the meagerness of the anticoagulant medication in the impediment and treatment of thromboembolic disorders, nevertheless enhancing the safety of treatment. In the United States, the LMWHs are presently acknowledged for the prophylaxis and treatment of deep venous thrombosis. The uses of LMWH are also being protracted for complementary signs for the management of unstable angina and myocardial infarction (Deepa & Varalakshmi, 2003).

7. ISOLATION OF LMWH SULFATE

Heparin and HS are linear, heterogeneously sulfated, polyanionic in natural anticoagulant. These polysaccharides surrounded by minor amounts of unsubstituted D-GlcN show considerable sequence heterogeneity, giving rise to very complex structures (Saravanan et al., 2010; Vidhyanandhini et al., 2014).

7.1. Chemical synthesis of low molecular heparin/HS

Universal approaches in carbohydrate synthesis rely on competent protecting cluster exploitations. LMWH and HS oligosaccharide synthesis takes place in three steps: (i) initiation, (ii) elongation, and (iii) termination. LMWH and HS oligosaccharide synthesis also requires addition of stereoselective glycosylation reactions. A good yield and high stereoselectivity in the coupling of chemical reactions are critical features of efficient oligosaccharide synthesis. Configurations of the oligosaccharides continued in regular-to-good yield (50–63%) and are followed by O-deacetylation, O-sulfonation, removal of the protecting groups, and N-sulfonation to afford the three target oligosaccharides. There are various factors can convince the stereoselectivity (α/β proportion) in glycosylations, among the most important are the nature of the leaving group at the anomeric position and the replacement at C-2 position of the glycosyl donor (Karst & Linhardt, 2003).

In 1993, van Boeckel and Petitou examined the synthesis of various heparin oliogosaccharide analogs and explained the protein interaction with heparin to AT III mechanisms. Several reports have resulted in an enhanced acquaintance of heparin structure–activity associations and preparation of potent, simplified heparin oligosaccharides analogs, and the high structural multiplicity of heparin/HS characterizes a major challenge for all carbohydrate chemists undertaking their synthesis. Consequently, the preparation of defined heparin/HS oligosaccharides requires the synthesis of tailor made monomers. Several carbohydrate chemists are involved in heparin/HS oligosaccharide synthesis over the past decade, and having made considerable progress in heparin/HS synthesis, head toward the understanding of biological activities. Modern technologies in carbohydrate chemistry have made heparin/HS oligosaccharides more accessible while still posing a momentous synthetic challenge. Novel ideas in the direction of the synthesis of larger and more complex heparin/HS oligosaccharides are essential. Attributable to their structural difficulty, chemical synthesis of each diverse nature of heparin/HS oligosaccharide still represents as a frightening mission (Basten et al., 1994).

7.2. Enzymatic synthesis of low molecular heparin/HSs

Nowadays, the methods available for pharmaceutical production of heparin from marine sources also include five basic steps: (i) preparation of marine tissue; (ii) extraction of heparin from marine tissue; (iii) crude heparin recovery; (iv) heparin purification; and (v) recovery of purified heparin. UFH is produced by the mast cells of most marine species. It is extracted from marine animals (vertebrates/invertebrates). The enzymes like alkylases, protease, chondroitin lyases, and endonucleases play a major role in enzymatic hydrolysis of LMWH/HS (Saravanan & Shanmugam, 2011) from marine scallop (*Amussium pleuronectus*).

Both heparin and HS consist of alternating chains of uronic acid and glucosamine, sulfated to varying degrees, and have a molecular weight range of 5–35 kDa, with a mean of about 12–14 kDa. LMWH is produced from UFH by managed depolymerization with chemical (nitrous acid or alkaline hydrolysis) or enzymatic (heparinases) process. Heparin and HS can be depolymerized enzymatically by using bacterial heparin lyases. Heparinase I acts on both HS and heparin, heparinase II acts only on heparin, and heparinase III is specific for its action on HS and is used to confirm the presence of HS and distinguish it from heparin (Le Brun & Linhardt, 2001). The chemical or enzymatic processes give different end products, but there is no confirmation that these differences in chemical structure influence the biological function. The biological activities of any LMWH are mostly determined by its molecular weight distribution. The products presently offered have a typical molecular weight between 4 and 6 kDa.

7.3. Chromatography separation of LMWH/HS

Chromatography is the only technique used to isolate an LMWH/HS from a crude or UFH, which mainly depends on the physical and/or chemical properties of the individual LMWH/HS. There is no particular or straightforward method to purify all kinds of LMWH/HS. Methods and circumstances used in the purification process of LMWH/HS may have an effect on the inactivation of another. The ultimate target also has to be considered when choosing purification method. The purity required depends on the intention for which the LMWH/HS is required. However, if the LMWH/HS is aimed for remedial apply, it must be extremely pure and purification must then be done in several successive steps. The objective of purification of LMWH/HS is not only removal of contaminants (CS, HA & DS), but also the net yield and activity of LMWH/HS, which is used for pharmacological application.

Generally, ion-exchange (Amberlite IRA 120/900, DEAE-Cellulose) and gel filtration (Sephadex-G 100/150) chromatography are used for separation of LMW-GAG/HS from marine mollusks by using various concentrations of NaCl (Saravanan et al., 2010; Saravanan & Shanmugam, 2010, 2011; Fig. 3.4). The elemental contents and anticoagulant activity of LMWH/HS obtained by chromatography are comparable to commercial source. In addition to this, SAX (strong anionic exchange resin) chromatography is also used as suitable stationary phase for separation of LMWH/HS (Warda et al., 2003). The HPLC that is used for protein purification is not

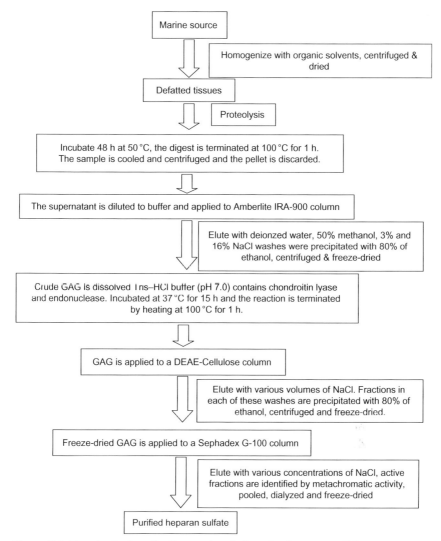

Figure 3.4 Flowchart of purification protocol of marine heparan sulfate.

suitable and there is no specific column for LMWH/HS separation, due to its structural variation. Selecting dialysis membrane cutoff size also plays a major role for separation of LMWH/HS from UFH. Further, the heparin is precipitated from mixtures or crude with 0.1–0.4 volumes of ethanol and, in particular, the fractions of 0.3 and 0.4 volumes are enriched with this GAG species. From 0.5 to 0.8 volumes, mixtures with different relative percentages of DS and CS are precipitated. In fact, at increasing volumes of ethanol (from 0.5 to 0.8), the amount of DS increases and that of CS decreases. Likewise, DS of about 95% purity can be obtained with 0.7–0.8 volumes of ethanol. With 1.0–2.0 volumes of ethanol, CS with high purity can be precipitated. The sequential precipitation of GAG species from mixtures using ethanol could be used to obtain high-purity heparin, DS (from 0.7 to 0.8 volumes) or CS with volumes greater than 1.0. These aptitudes of ethanol are utilized to purify heparin and CS from a bovine lung (Volpi, 1996). In the early days of carbohydrate chemistry, the only practical method to separate different types of carbohydrate is by taking improvement of their relative solubility. Fraction of a LMWH/HS is caused to precipitate through the alteration of some properties of the solvent, for example, addition of salts, organic solvents preparation, distilled water, or altering the pH or temperature. Fractional precipitation is still commonly used for separation of gross impurities, PG, and nucleic acids.

8. CONCLUSION

Marine products referring here to miniature molecules fashioned by living organisms, particularly sulfated polysaccharides such as heparin, HS, CS, and DS are an exceptional yet currently underutilized foundation of chemical biodiversity for drug discovery. Bioactive substances isolated from marine plants, fungi, and bacteria have given rise to a wide range of creature therapeutics, and abundant others are effective methods in marine biology, biochemistry, and pharmacology. Several challenges unique to marine products have contributed to this recent trend as well, including the difficulties associated with isolating pure LMWH/HS from crude GAGs, identifying their mechanism of action and synthesizing these often times highly complex structures of heparin.

The anticoagulant activity and content of elements of heparin/HS from marine mollusk are similar to commercial pharmaceutical heparin. The degree and percentage of sulfate content is also high in marine mollusks, which ensure the anticoagulant and other biochemical activities. Chromatography is the only economically viable techniques for separation of LMWH/HS from marine organisms when compared to enzyme hydrolysis. Since very few studies are carried out hitherto the genomics and sequencing of heparin from marine organisms and yet to study the binding interaction of LMWH/HS to proteins (fibroblast growth factor, interleukin etc.) to ascertain molecular mechanisms of marine LMWH/HS. When we conquer, these intentions do not necessitate perturbing about the diversity of marine organisms. Although this testimony of accomplishment of LMWH/HS, the field of research in marine product has non-the-less witnessed a noteworthy turn down in the past two decades, due in great gauge to the materialization of target corroboration, combinatorial carbohydrate chemistry and high-throughput broadcast as a new exemplar for drug discovery in future.

REFERENCES

Basten, J. E. M., van Boeckel, C. A. A., Jaurand, G., Petitou, M., Soijker, N. M., & Westerduin, P. (1994). Synthesis of a dermatan sulphate-like hexasaccharide with a non-glycosamino-glycan structure. *Bioorganic & Medicinal Chemistry Letters, 4*, 893–898.

Best, C. H. (1959). Preparation of heparin and its use in the first clinical cases. *Circulation, 19*(1), 79–86.

Bhattacharya, A. (2008). *Flask synthesis promises untainted heparin.* Chemistry World. Royal Society of Chemistry.

Deepa, P. R., & Varalakshmi, P. (2003). Protective effect low molecular weight heparin on oxidative injury and cellular abnormalities in adriamycin-induced cardiac and hepatic toxicity. *Chemico-Biological Interactions, 146*, 201–210.

Gallagher, J. T., Turnbull, J. E., & Lyon, M. (1992). Patterns of sulphation in heparan sulphate: Polymorphism based on a common structural theme. *International Journal of Biochemistry, 24*, 553–560.

Gallagher, J. T., & Walker, A. (1985). Molecular distinctions between heparan sulphate and heparin. Analysis of sulphation patterns indicates that heparan sulphate and heparin are separate families of N-sulphated polysaccharides. *Biochemical Journal, 230*(3), 665–674.

Gordon, P. A., & Strickland, S. (1989). Anticoagulant low molecular weight heparin does not enhance the activation of plasminogen by tissue plasminogen activator. *The Journal of Biological Chemistry, 264*(15), 15177–15181.

Guo, X., Condra, M., Kimura, K., Berth, G., Dautzenberg, H., & Dubin, P. L. (2003). Determination of molecular weight of heparin by size exclusion chromatography with universal calibration. *Analytical Biochemistry, 312*, 33–39.

Karst, N. A., & Linhardt, R. J. (2003). Recent chemical and enzymatic approaches to the synthesis of glycosaminoglycan oligosaccharides. *Current Medicinal Chemistry, 10*, 1993–2031.

Le Brun, L. A., & Linhardt, R. J. (2001). Degradation of heparan sulfate with heparin lyases. In R. V. Iozzo & N. J. Totowa (Eds.), *Methods in molecular biology, proteoglycan protocols* (pp. 351–359). Humana press. Current methods and applications, Chap, 35, 171.

Lindahl, U., Lidholt, K., Spillmann, D., & Kjellen, L. (1994). More to "heparin" than anticoagulation. *Thrombosis Research, 75*, 1–32.

Linhardt, R. J. (1991). Heparin: An important drug enters its seventh decade. *Chemistry & Industry, 2*, 45–50.

Linhardt, R. J., & Claude, S. (2003). Hudson award address in carbohydrate chemistry. Heparin: Structure and activity. *Journal of Medicinal Chemistry, 46*, 2551–2564.

Linhardt, R. J., & Gunay, N. S. (1999). Production and chemical processing of low molecular weight heparins. *Seminars on Thrombosis and Hemostasis, 25*(3), 5–16.

Linhardt, R. J., & Toida, T. (1997). Heparin oligosaccharides: New analogues development and applications. In Z. J. Witczak, & K. A. Nieforth (Eds.), *Carbohydrates in drug design* (pp. 277–341). New York: Marcel Dekker.

Linhardt, R. J., & Toida, T. (2004). Role of glycosaminoglycans in cellular communication. *Accounts of Chemical Research, 37*, 431–438.

Mc Lean, J. (1916). The thromboplastic action of cephalin. *American Journal of Physiology, 41,* 250–257.
Monique, S., Kher, A., & Toulemonde, F. (Eds.). (1999). *Low molecular weight heparin therapy—An evaluation of clinical trials evidence* (p. 12). New York: Marcel Dekker.
Mousa, S. A. (2007). *Methods in molecular medicine. Anticoagulants, antiplatelets, and thrombolytics.* Totowa, NJ: Humana Press Inc, 93, 1–7. P 292.
Ori, A., Wilkinson, M. C., & Fernig, D. G. (2011). A systems biology approach for the investigation of the heparin/heparan sulfate interactome. *The Journal of Biological Chemistry, 286*(22), 19892–19904.
Pavao, M. S. G., & Mourao, P. A. S. (2012). Challenges for heparin production: Artificial synthesis or alternative natural sources? *Glycobiology Insights, 3,* 1–6.
Sampaio, L. O., Tersariol, I. L. S., Lopes, C. C., Boucas, R. I., Nascimento, F. D., Rocha, H. A. O., et al. (2006). Heparins and heparan sulfates. Structure, distribution and protein interactions. In H. Verli (Ed.), *Insights into carbohydrate structure and biological function* (pp. 1–19). Trivandrum, India: Transworld Research Network.
Saravanan, R. (2010). *Bioactive compounds from marine bivalve mollusk.* PhD Thesis, Parangipettai, India: Annamalai University, pp. 110–120.
Saravanan, R., & Shanmugam, A. (2010). Isolation and characterization of low molecular weight glycosaminoglycans from marine mollusc *Amussium pleuronectus* (Linne) using Chromatography. *Applied Biochemistry and Biotechnology, 160,* 791–799.
Saravanan, R., & Shanmugam, A. (2011). Is isolation and characterization of heparan sulfate from marine scallop *Amussium pleuronectus* (Linne) an alternative source of heparin?!! *Carbohydrate Polymers, 86*(2), 1082–1084.
Saravanan, R., Vairamani, S., & Shanmugam, A. (2010). Glycosaminoglycans from marine clam *Meretrix meretrix* (Linne.) are an anticoagulant. *Preparative Biochemistry and Biotechnology, 40,* 305–315.
Schonberger, L. B. (1998). New variant Creutzfeldt-Jakob disease and bovine spongiform encephalopathy. *Infectious Disease Clinics of North America, 12,* 111–121.
Stringer, S. E., & Gallagher, J. T. (1997). Heparan sulphate. *International Journal of Biochemistry & Cell Biology, 29*(5), 709–714.
Takats, G. D. (1950). The subcutaneous use of heparin. *Circulation, 2,* 837–844.
Toida, T., Yoshida, H., Toyoda, H., Koshiishi, I., Imanari, T., Hileman, R. E., et al. (1997). Structural differences and the presence of unsubstituted amino groups in heparan sulphates from different tissues and species. *Biochemical Journal, 322*(2), 499–506.
Vidhyanandhini, R., Saravanan, R., Vairamani, S., & Shanmugam, A. (2014). The anticoagulant activity and structural characterization of fractionated and purified glycosaminoglycans from venerid clam *Meretrix casta* (Chemnitz). *Journal of Liquid Chromatography & Related Technologies, 37,* 917–929.
Volpi, N. (1996). Purification of heparin, dermatan sulfate and chondroitin sulfate from mixtures by sequential precipitation with various organic solvents. *Journal of Chromatography B, 685,* 27–34.
Vyas, N., Goswami, D., Manonmani, A., Sharma, P., Ranganath, H. A., VijayRaghavan, K., et al. (2008). Nanoscale organization of hedgehog is essential for long-range signaling. *Cell, 133,* 1214–1227.
Warda, M., Gouda, E. M., Toida, T., Chi, L., & Linhardt, R. J. (2003). Isolation and characterization of raw heparin from dromedary intestine: Evaluation of a new source of pharmaceutical heparin. *Comparative Biochemistry and Physiology Part C, 136,* 357–365.
Wassener, B. (2010). In China, Strong debut for supplier of heparin. *New York Times.*
Yu, S. R., Burkhardt, M., Nowak, M., Ries, J., Petra´sek, Z., Scholpp, S., et al. (2009). Fgf8 morphogen gradient forms by a source-sink mechanism with freely diffusing molecules. *Nature, 461,* 533–536.

CHAPTER FOUR

Isolation and Characterization of Hyaluronic Acid from Marine Organisms

Sadhasivam Giji, Muthuvel Arumugam[1]

Centre of Advanced Study in Marine Biology, Faculty of Marine Sciences, Annamalai University, Parangipettai, Tamil Nadu, India
[1]Corresponding author: e-mail address: mamnplab@gmail.com

Contents

1. Introduction — 62
2. Targeted Sources for HA in the Past and Present Era — 62
3. Structure of Hyaluronic Acid — 64
4. Isolation Methods — 64
    4.1 Extraction by enzyme digestion method — 66
    4.2 Extraction with organic solvents and sodium acetate — 66
    4.3 Microbial production — 67
    4.4 Supplementary methods — 67
5. Characterization of Hyaluronic Acid — 67
    5.1 Electrophoretic analysis — 68
    5.2 Spectroscopic investigation — 68
    5.3 Compositional analysis — 70
    5.4 Determination of molecular weight and viscosity — 70
6. Conclusion — 71
Acknowledgments — 73
References — 73

Abstract

Hyaluronic acid (HA) being a viscous slippery substance is a multifunctional glue with immense therapeutic applications such as ophthalmic surgery, orthopedic surgery and rheumatology, drug delivery systems, pulmonary pathology, joint pathologies, and tissue engineering. Although HA has been isolated from terrestrial origin (human umbilical cord, rooster comb, bacterial sources, etc.) so far, the increasing interest on this polysaccharide significantly aroused the alternative search from marine sources since it is at the preliminary level. Enthrallingly, marine environments are considered more biologically diverse than terrestrial environments. Although numerous methods have been described for the extraction and purification of HA, the hitch on the isolation methods which greatly influences the yield as well as the molecular weight of the polymer still exists. Adaptation of suitable method is essential in this venture. Stimulated by the developed technology, to sketch the steps involved in isolation and analytical techniques for characterization of this polymer, a brief report on the concerned approach has been reviewed.

1. INTRODUCTION

The hyaluronic acid (HA), most versatile compound also known as hyaluronan, classified as a glycosaminoglycan (commonly known as a GAG), is an anionic, linear, nonsulfated polysaccharide. HA is a mega Dalton molecule, with a typical molecular weight between $\sim 2 \times 10^5$ and $\sim 10 \times 10^7$ Da and an extended length of 2–25 μm. However, other GAGs are relatively smaller in size between 50 kDa, commonly 15–20 kDa with a short-chain length. Also, HA is synthesized at the inner face of the plasma membrane as a free linear polymer without any protein core, while other GAGs are synthesized by resident Golgi enzymes and covalently attached to core proteins (Laurent & Fraser, 1992; Lee & Spicer, 2000; Toole, 2004). HA is almost omnipresent (albeit in relatively small amounts) in the human body, and in other vertebrates, the highest amounts of HA are found in the umbilical cord, nasal cartilage, vitrium cutis, lymph of the thorax, plasma, the synovial fluid, and connective tissue such as the synovial membrane, where it is responsible for normal water retention and lubrication of the joint. HA possesses significant structural, rheological, physiological, and biological functions with distinctive moisturizing retention ability and viscoelasticity, coupled with its lack of immunogenicity and toxicity; hence, HA finds various applications in the cosmetic, biomedical, and food industries (Chong, Blank, Mc laaughlin, & Nielsen, 2005). This offers great promise in various fields of medicine, and worldwide market for HA is estimated to cross $2.5 billion by 2017. Consequently, the HA-derived therapeutics emphasizes the impetus for the development of biotechnological and chemical processes for optimization of the production of HA-based drugs. Hence, the methods involved in isolation and characterization of HA have been described briefly.

2. TARGETED SOURCES FOR HA IN THE PAST AND PRESENT ERA

In terms of evolution and biodiversity, the sea appears to be superior to the terrestrial ecosystem; marine species comprise approximately a half of the total biodiversity, thus offering a vast source to discover useful therapeutics.

The therapeutic potential of natural bioactive compounds such as polysaccharides, especially glycosaminoglycans, is now well documented, and this activity combined with natural biodiversity will allow the development of a new generation of therapeutics (Senni et al., 2011). Glycosaminoglycans abound in vertebrate tissues, and invertebrate species are rich source of polysaccharides with novel structures (Albano & Mourao, 1986; Cassaro & Dietrich, 1977; Vieira & Mourao, 1988). Figure 4.1 represents the distribution of GAGs in marine vertebrates and invertebrates.

Sulfated polysaccharides including chondroitin sulfate (Nader et al., 1984), dermatan sulfate (Cassaro & Dietrich, 1977), heparan sulfate (Oliveira, Chavante, Santos, Dietrich, & Nader, 1994), heparin (Dietrich et al., 1985), acharan sulfate (Kim et al., 1996) have also been isolated and characterized from different marine sources. But, work on isolation of nonsulfated HA from marine source is very few in spite of its various fundamental biological properties and functions.

HA occurs primarily in the ECM and pericellular matrix, although recently it has been shown to be present intracellularly (Evanko & Wight, 2001). It is therefore one of the largest components of the ECM, whose structure appears identical throughout phyla and species as diverse as Pseudomonas slime, Ascaris worms, and mammals such as the rat, rabbit, and

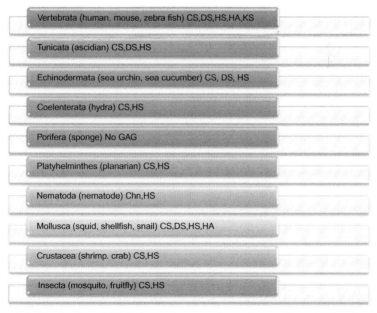

Figure 4.1 Distribution of GAGs in marine organisms. *Source: Yamada, Sugahara, and Ozbek (2011).*

human. Equally, its presence has been documented in tissues as diverse as the skin, aorta, cartilage, and brain (Price, Berry, & Navsaria, 2007). Thus, HA has been obtained from various natural terrestrial sources for clinical applications and research so far (Table 4.1).

Recently, in the last decade, prospecting on marine organism for HA was increased. Volpi (2003) has isolated antiproliferative potential low-molecular-weight (200 kDa) HA from marine bivalve *Mytilus galloprovincialis*. Recently, we have isolated high-molecular-weight HA with similar activity from Stingray (*Aetobatus narinari*) liver (Giji, Arumugam, Abirami, & Balasubramanian, 2012) and low-molecular-weight antioxidant potent HA from marine bivalve *Amussium pleuronectus* (Kanchana, Arumugam, Giji, & Balasubramanian, 2013). HA with high purity (more than 99.5%), useful for clinical and cosmetic applications, is obtained by means of low-cost process using eyeball (Murado, Montemayor, Cabo, Vazquez, & Gonzalez, 2012). Vazquez, Montemayor, Fraguas, and Murado (2010) have used Mussel-processing wastewater as a sugar source and tuna peptone from viscera residue as a protein substrate for the production of HA by *Streptococcus zooepidemicus* in batch fermentation. Though the source target for isolation of HA is abundant in the marine environment, the work on this particular component is countable.

3. STRUCTURE OF HYALURONIC ACID

Hyaluronan occurs mainly in the extracellular matrix and is thought to take part in many biological processes and functions. This high-molecular-weight biopolymer is considered to be energetically stable, in part because of the stereochemistry of its component disaccharides. Bulky groups on each sugar molecule are in sterically favored positions, whereas the smaller hydrogens assume the less-favorable axial positions. The structural difference between the sulfated and nonsulfated GAGs has been shown in Fig. 4.2.

4. ISOLATION METHODS

The breaking down of the cellular structures of the tissue and the extraction of hyaluronan from other polysaccharides complex is achieved by crushing followed by action of organic solvents, enzymes, and detergents. Methods based on these strategies for isolation of HA from marine sources have been reviewed here.

Table 4.1 List of terrestrial sources used for HA isolation

S. no	Animal	Tissue/body fluid	Literature sources
Terrestrial sources			
1.	Rooster	Rooster comb	Kang, Kim, Heo, Park, and Lee (2010)
2.	Human	Umbilical cord	Hadidian and Pirie (1948)
		Joint synovial fluid	Balazs, Watson, Duff, and Roseman (2005)
		Vitreous body	Nishikawa and Tamai (1996)
		Dermis	Postlethwaite et al. (1989)
		Epidermis	Akiyama, Saito, Qiu, Toida, and Imanari (1994)
		Thoracic lymph	Pethrick, Ballada, and Zaikov (2007)
		Urine	Toyoda et al. (1991)
		Serum	Deutsch (1957)
3.	Rat	Lung	Nettelbladt, Bergh, Schenholm, Tengblad, and Hallgren (1989)
		Kidney	Hallgren, Gerdin, and Tufveson (1990)
		Brain	Bignami and Asher (1992)
		Liver	Fraser, Alcorn, Laurent, Robinson, and Ryan (1985)
		Aqueous humor	Kuhlman and Kaufman (1960)
4.	Cow	Bovine nasal cartilage	Cleland and Sherblom (1977)
5.	Sheep	Synovial fluid	Fraser, Kimpton, Pierscionek, and Cahill (1993)
		Medulla cortex	Dicker and Franklin (1966)
		Lung	Lebel, Smith, Risberg, Laurent, and Gerdin (1988)
6.	Rabbit	Renal papillae	Farber and VagPraag (1970)
		Kidney	Dicker and Franklin (1966)
		Vitreous body	Necas, Bartosikova, Brauner, and Kolar (2008)
		Renal cortex	Dwyer et al. (2000)
		Muscle	Necas et al. (2008)
		Liver	Takagaki, Nakamura, Majima, and Endo (1988)
		Aqueous humor	Laurent and Fraser (1983)
7.	Bacteria	Streptococci sp.	Seastone (1939)

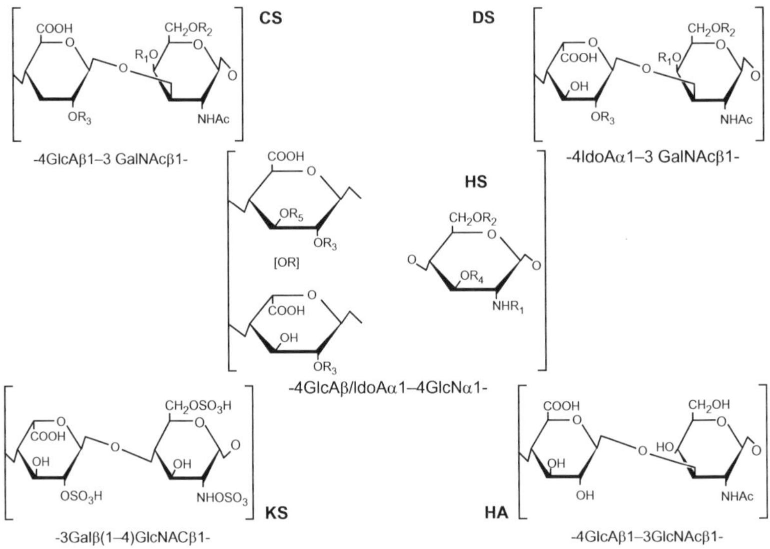

Figure 4.2 Structure of sulfated and nonsulfated (HA) GAGs.

4.1. Extraction by enzyme digestion method

The tissue was defatted with acetone and dried at 60 °C for 24 h. The dried pellet was solubilized in 100 mM sodium acetate buffer, pH 5.5, containing 5 mM EDTA and cysteine. 100 mg of papain was added per gram of tissue, and the solution was incubated for 24 h at 60 °C in a stirrer. After boiling for 10 min, the mixture was centrifuged and three volumes of ethanol saturated with sodium acetate were added to the supernatant and stored at 4 °C for 24 h. The precipitate was recovered by centrifugation and dried at 60 °C (Volpi, 2003). The tissues can also be digested with pepsin (Bychkov & Kolesnikova, 1969), pronase (Sven & Zhag, 1986), and trypsin (Ogston & Sherman, 1958).

4.2. Extraction with organic solvents and sodium acetate

The tissue was homogenized with acetone and incubated for 24 h in refrigerator. After 24 h, the acetone was squeezed from the tissue material. This step was repeated 10 times in 24-h intervals. This material was extracted 10 times successively with a 5% solution of sodium acetate; each time the viscous fluid was squeezed through several layers of cheesecloth. 1.5 volumes of ethyl alcohol were added to the aqueous extracts. The precipitates formed were pooled, centrifuged, redissolved in 5% sodium acetate solution, and recentrifuged. Protein was removed from the supernatant solution by shaking it with chloroform four times and then with a chloroform–amyl alcohol (1:4 parts to 1:2 parts) mixture several times until a gel no longer formed. The final solution was dialyzed; sodium acetate crystals were added to make a 5% solution. Following acidification to pH 4.0, the solution was precipitated with ethyl alcohol and the precipitate was desiccated in vacuum over calcium chloride. The final dried material will be pure white and fibrous in appearance (Boas, 1949).

4.3. Microbial production

HA can be prepared in high yield from bacterial sources by fermenting the bacteria under anaerobic conditions in an enriched growth medium containing glucose, glycogen, yeast extract, tryptone, KH_2PO_4, K_2HPO_4, $MgSO_4 \cdot 7H_2O$, $(NH_4)_2SO_4$, and polystyrene (Vazquez et al., 2010). HA was precipitated by mixing with three volumes of ethanol and centrifuged at $5000 \times g$ for 10 min. The sediment was resuspended with one volume of NaCl (1.5 M) and three volumes of ethanol and precipitated by centrifugation at $5000 \times g$ for 10 min. Finally, this last sediment was redissolved in distilled water for HA analysis (Vazquez, Montemayor, Fraguas, & Murado, 2009). The yield, purity, and low cost of the hyaluronic acid produced by the bacterial sources also permit it to be used in ways not previously described or contemplate for hyaluronic acid obtained from mammalian or low yield bacterial sources.

4.4. Supplementary methods

The separation of the acid mucopolysaccharides into sulfated and non-sulfated fractions was achieved with a cetylpyridium complex (Scott, 1960), and final purification of the fractions was obtained by use of a DEAE-Sephadex anion exchanger A-25 (Schmidt, 1962). The centrifugal precipitation chromatography (Shinomiya, Kabasawa, Toida, Imanari, & Ito, 2001), electrodeposition, and ultrafiltration–diafiltration (Murado et al., 2012) are the other means of separation of HA fragments.

5. CHARACTERIZATION OF HYALURONIC ACID

Several analytical approaches have been developed to evaluate the species and quantity of various GAGs. Furthermore, analytical techniques are of paramount importance for evaluating the purity of a single GAG species

(Volpi & Maccari, 2006). Various methods have been described for the separation of hyaluronan oligomers. Most of these involve the digestion of polymeric HA with the endohydrolase, testicular hyaluronidase, followed by purification by size exclusion chromatography (Lesley, Hascall, Tammi, & Hyman, 2000; Tammi et al., 1998), ion exchange chromatography (Almond, Brass, & Sheehan, 1998; Holmbeck & Lerner, 1993; Toffanin et al., 1993), and reversed-phase ion pair high-performance liquid chromatography (Cramer & Bailey, 1991).

5.1. Electrophoretic analysis

The electromigration procedures developed to analyze and characterize complex polysaccharides including electrophoresis on cellulose acetate (Wessler, 1968) and nitrocellulose membrane (Volpi, 1996), agarose gel electrophoresis (Dietrich & Dietrich, 1976; Dietrich, McDuffie, & Sampaio, 1977), electrophoresis on polyacrylamide gel (Edens et al., 1992; Hakim & Linhardt, 1990; Min & Cowman, 1986; Rice, Rottink, & Linhardt, 1987), high-performance capillary electrophoresis (Maccari & Volpi, 2003; Volpi, 2004a, 2004b), and fluorophore-assisted carbohydrate electrophoresis (Calabro et al., 2001; Volpi & Maccari, 2005) analyses were critically important in HA evaluation. These methods allow on direct detection or prederivatization for both qualitative and quantitative analysis with a high level of sensitivity (Volpi & Maccari, 2006).

5.2. Spectroscopic investigation

HA is a multinegative biopolymer and is known to interact with many cationic dyes forming stable complexes. Based on this property, the methods using Stains-all are reported for the determination of HA in solutions, human tissues, biological fluids (Beighton, Pahuja, Gray, & Edstrom, 1977; Homer, Denbow, Whiley, & Beighton, 1993; Kay, Walwick, & Gilford, 1954; Pryce-Jones & Lannigan, 1979). Fagnola, Pagani, Maffioletti, Tavazzi, and Papagni (2009) introduced a spectroscopic method for the determination of HA in solutions based on the formation of complex between HA and Stains-all. The HA dye complex will form peak at 640 nm, while the free dye forms peak at 520 nm. This is one of the useful methods of determining the presence of HA in the isolated samples.

The vibrational spectrum of a molecule is considered to be a unique physical property and is characteristic of the molecule. The infrared spectrum can be used as a fingerprint for identification by the comparison of the spectrum from an unknown with previously recorded reference spectra. Over the years, much has been published in terms of the fundamental absorption frequencies which are the key to unlock the structure–spectral relationships of the associated molecular vibrations (Coates, 2000). The presence of hydroxy group H-bonded —OH stretch and N—H stretch, methylene C—H asym/sym stretch, CH stretching, C═O carboxyl amid I, CH2, CH3, C—O—H deformation, C—O with C—O combination, aromatic primary amine CN stretch, C—O—C, C—O, and C—O—H stretching in the sample can be identified and aligned with standard HA through FTIR analysis (Alkrad, Mrestani, Stroehl, Wartewig, & Neubert, 2003; Giji et al., 2012; Kanchana et al., 2013). Similarly, through the analysis of proton and carbon NMR, the presence of anomeric protons, β-D-glucopyranosiduronate unit and 2-acetamido-deoxy-β-D-glucopyranoside unit of HA can be revealed (Bociek, Darke, Welt, & Rees, 1980; Tommeraas & Melander, 2008). The various spectra of anionic polymer HA have been shown in Fig. 4.3.

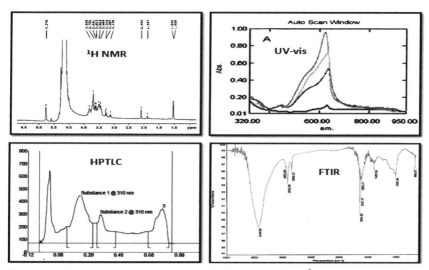

Figure 4.3 Spectroscopic investigation of hyaluronic acid. (A) ^{1}H NMR spectrum of HA from marine vertebrate (Giji et al., 2012). (B) ^{13}C NMR spectrum of HA from marine bivalve (Volpi et al., 2003). (C) FTIR spectrum of standard and marine bivalve *A. pleuronectus* (Kanchana et al., 2013).

5.3. Compositional analysis

Uronic acid and N-acteyl glucosamine are the components of the repeating unit of HA. The quantification of these subunits is essential in characterization of HA. The carbazole method developed by Bitter and Muir (1962) was the commonly used method for uronic acid estimation. Uronic acids can be quantified either by a colorimetric determination with concentrated sulfuric acid and carbazole or by gas chromatography after methanolysis and subsequent acetylation. Since both the methods suffer from incomplete hydrolysis, an unavoidable degradation of the products has to be analyzed. Li, Kisara, Danielsson, and Lindstrom (2007) have developed an improved method for quantification of uronic acid by modifying the fundamental chemistry involved in the above said methods. The N-acetyl glucosamine content of the sample was determined by following the method of Reissig, Strominger, and Leloir (1955).

Besides this, di-, tetra-, and hexasaccharides of HA were analyzed earlier by strong anion exchange HPLC separation (Volpi, 2003). Later, Zhang et al. (2007) have developed thin-layer chromatography method for rapid and semiquantitative analysis of glycosaminoglycans. It was also evidenced that the absence of sulfation in HA oligosachharides resulted in both the enhanced migration and resolution of these saccharides when compared to other sulfated GAGs.

5.4. Determination of molecular weight and viscosity

The applications of HA depend on its molecular weight, which is an important quality parameter for charactering commercial HA products (Liu, Liu, Li, Du, & Chen, 2011). HA in aqueous solution has been reported to undergo a transition from Newtonian to non-Newtonian characteristics with increasing molecular weight, concentration, or shear rate. In addition, the higher the molecular weight and concentration of HA, the higher the viscoelasticity the solutions possess. The especially high-molecular-weight hyaluronates, because of their high viscosity in solution, probably exhibit intermolecular interaction. Hence, the evaluation of molecular weight and viscosity is significant from application point of HA.

The molecular weight of HA can be measured either through high-performance size exclusion chromatography (Orvisky et al., 1992) or through agarose gel electrophoresis (Cowman et al., 2011). Yeung and Marecak (1999) involved the combined use of aqueous gel filtration chromatography (GFC) with matrix-assisted laser desorption ionization mass spectrometry (MALDI-MS) for molecular weight determination of HA and further reported that the GFC–MALDI-MS approach is a reliable method for the molecular weight characterization of polydisperse polysaccharides for which suitable calibration standards are unavailable for conventional GFC analysis. Furthermore, Dong et al. (2010) demonstrated that NIR spectroscopy was superior to conventional methods for the fast determination of molecular weight of hyaluronic acid.

The extraordinary rheological properties of hyaluronan solutions make them ideal as lubricants. HA solutions are characteristically viscoelastic and pseudoplastic. At higher concentrations, solutions have an extremely high but shear-dependent viscosity. This rheology is found even in very dilute solutions of the polymer where very viscous gels are formed. In solutions, the hyaluronan polymer chain takes on the form of an expanded, random coil. These chains entangle with each other at very low concentrations, which may contribute to the usual rheological properties (Saranraj & Naidu, 2013). The ratio of flow time of sample to that of solvent was taken as the relative viscosity. Viscosity can be analyzed in Ostwald viscometer following the method of Hadidian and Pirie (1948). In order to elucidate the relationship between conformation and molecular weight of glycosaminoglycans, it is important to clarify the significance of the constants of the viscosity equations, k' and k''.

6. CONCLUSION

The expanding application of HA in various fields of medicine had shed light in the universal target toward this polymer since 1934. Apart from this, the worldwide market value for HA shoots up every year. Though several tons HA per year have been produced by attenuated microbes, the risk of mutation, coproduction of various toxins, etc., hinders their application in clinical practice. Consequently, the hunt on source target for mass scale isolation of HA is also increasing. In this prospect, the sea covering around 70% of our planet's surface represents the highest percentage of the living biosphere and is the greatest source of biodiversity. Previously, in the course of our investigation on isolation of HA from vertebrates and invertebrates, we have reported in relatively high yield of both low- and high-molecular-weight nonsulfated GAGS with antioxidant and antiproliferation activity. It is very apparent that Ocean offers an enormous diversity of organisms for isolation and production of HA economically, wherein the plethora of marine organisms will advantageously replace the terrestrial sources for isolation of HA in the near years. Henceforth, we have given an overview on strategies involved in the isolation and characterization of HA from marine organisms (Fig. 4.4).

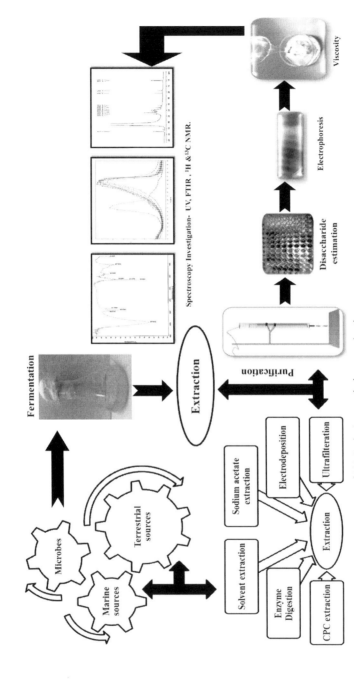

Figure 4.4 Schematic representation of HA isolation and characterization.

ACKNOWLEDGMENTS
The authors were grateful to DBT (G4/8855) for their financial assistance. We thank Annamalai University authorities for providing us necessary facilities.

REFERENCES
Akiyama, H., Saito, M., Qiu, G., Toida, T., & Imanari, T. (1994). Analytical studies on hyaluronic acid synthesis by normal human epidermal keratinocytes cultured in a serum-free medium. *Biological and Pharmaceutical Bulletin, 17*(3), 361–364.

Albano, R. M., & Mourao, P. A. S. (1986). Isolation, fractionation, and preliminary characterization of a novel class of sulphated glycans from the tunic of *Styelaplicata* (ChordataTunicata). *Journal of Biological Chemistry, 261*, 758–765.

Alkrad, J. A., Mrestani, Y., Stroehl, D., Wartewig, S., & Neubert, R. (2003). Characterization of enzymatically digested hyaluronic acid using NMR, Raman, IR and UV-Vis spectroscopies. *Journal of Pharmaceutical and Biomedical Analysis, 31*, 545–550.

Almond, A., Brass, A., & Sheehan, J. K. (1998). Deducing polymeric structure from aqueous molecular dynamics simulations of oligosaccharides: Predictions from simulations of hyaluronan tetrasaccharides compared with hydrodynamic and X-ray fibre diffraction data. *Journal of Molecular Biology, 284*(5), 1425–1437.

Balazs, E. A., Watson, D., Duff, I. F., & Roseman, S. (2005). Hyaluronic acid in synovial fluid. I. Molecular parameters of hyaluronic acid in normal and arthritic human fluids. *Arthritis and Rheumatism, 10*(4), 357–376.

Beighton, D., Pahuja, S., Gray, E., & Edstrom, D. (1977). A sensitive method for assay of hyaluronidase activity. *Analytical Biochemistry, 79*, 431–437.

Bignami, A., & Asher, R. (1992). Some observations on the localization of hyaluronic acid in adult, newborn and embryonal rat brain. *International Journal of Developmental Neuroscience, 10*(1), 45–57.

Bitter, T., & Muir, H. M. (1962). A modified uronic acid carbazole reaction. *Analytical Biochemistry, 4*, 330–334.

Boas, N. F. (1949). Isolation of hyaluronic acid from the cock's comb. *Journal of Biological Chemistry, 181*, 573–575.

Bociek, S. M., Darke, A. H., Welt, D., & Rees, D. A. (1980). The ^{13}C-NMR spectra of hyaluronate and chondroitin sulphates. Further evidence on an alkali-induced conformation change. *European Journal of Biochemistry, 109*(2), 447–456.

Bychkov, S. M., & Kolesnikova, M. F. (1969). Investigation of highly purified preparations of hyaluronic acid. *Biokhimiya, 34*(1), 204–208.

Calabro, A., Midura, R., Wang, A., West, L., Plaas, A., & Hascall, V. C. (2001). Fluorophore-assisted carbohydrate electrophoresis (FACE) of glycosaminoglycans. *Osteoarthritis and Cartilage, 9*(A), 16–22.

Cassaro, C. M., & Dietrich, C. P. (1977). Distribution of sulphated mucopolysaccharides in invertebrates. *The Journal Biological Chemistry, 252*, 2254–2261.

Chong, F. B., Blank, L. M., Mc laaughlin, R., & Nielsen, L. K. (2005). Microbial hyaluronic acid production. *Applied Microbiology and Biotechnology, 66*, 341–351.

Cleland, R. L., & Sherblom, A. P. (1977). Isolation and physical characterization of hyaluronic acid prepared from bovine nasal septum by cetylpyridinium chloride precipitation. *Journal of Biological Chemistry, 252*, 420–426.

Coates, John. (2000). Interpretation of infrared spectra, a practical approach. In R. A. Meyers (Ed.), *Encyclopedia of analytical chemistry* (pp. 10815–10837). Chichester: John Wiley & Sons Ltd.

Cowman, M. K., Chen, C. C., Pandya, M., Yuan, H., Ramkishun, D., Lobello, J., et al. (2011). Improved agarose gel electrophoresis method and molecular mass calculation for high molecular mass hyaluronan. *Analytical Biochemistry, 417*(1), 50–56.

Cramer, J. A., & Bailey, L. C. (1991). A reversed-phase ion-pair high-performance liquid chromatography method for bovine testicular hyaluronidase digests using post column derivatization with 2-cyanoacetamide and ultraviolet detection. *Analytical Biochemistry, 196*, 183–191.

Deutsch, H. F. (1957). Some properties of a human serum hyaluronic acid. *Journal of Biological Chemistry, 224*, 767–774.

Dicker, S. E., & Franklin, C. S. (1966). The isolation of hyaluronic acid and chondroitin sulphate from kidneys and their reaction with urinary hyaluronidase. *Journal of Physiology, 186*(1), 110–120.

Dietrich, C. P., de Paiva, J. F., Moraes, C. T., Takahashi, H. K., Porcionatto, M. A., & Nader, H. B. (1985). Isolation and characterization of a heparin with high anticoagulant activity from *Anomalocardia brasiliana*. *Biochimica et Biophysica Acta, 843*, 1–7.

Dietrich, C. P., & Dietrich, S. M. C. (1976). Electrophoretic behavior of acidic mucopolysaccharides in diamine buffers. *Analytical Biochemistry, 70*, 645–647.

Dietrich, C. P., McDuffie, N. M., & Sampaio, L. O. (1977). Identification of acidic mucopolysaccharides by agarose gel electrophoresis. *Journal of Chromatography, 130*, 299–304.

Dong, Q., Zang, H., Liu, A., Yang, G., Sun, C., Sui, L., et al. (2010). Determination of molecular weight of hyaluronic acid by near-infrared spectroscopy. *Journal of Pharmaceutical and Biomedical Analysis, 53*(3), 274–278.

Dwyer, T. M., Banks, S. A., Alonso-Galicia, M., Cockrell, K., Carroll, J. F., Bigler, S. A., et al. (2000). Distribution of renal medullary hyaluronan in lean and obese rabbits. *Kidney International, 58*, 721–729.

Edens, R. E., Al-Hakim, A., Weiler, J. M., Rethwisch, D. G., Fareed, J., & Linhardt, R. J. (1992). Gradient polyacrylamide gel electrophoresis for determination of molecular weights of heparin preparations and low-molecular-weight heparin derivatives. *Journal of Pharmaceutical Sciences, 81*(8), 823–827.

Evanko, S., & Wight, T. (2001). Intracellular hyaluronan. In *Hyaluronan: Synthesis, function, catabolism*. Available at, http://www.glycoforum.gr.jp/science/hyaluronan/HA20/HA20E.html, Cited 30 July 2001.

Fagnola, M., Pagani, M. P., Maffioletti, S., Tavazzi, S., & Papagni, A. (2009). Hyaluronic acid in hydrophilic contact lenses: Spectroscopic investigation of the content and release in solution. *Contact Lens & Anterior Eye, 32*(3), 108–112.

Farber, S. J., & VagPraag, D. (1970). Composition of glycosaminoglycans mucopolysaccharides in rabbit renal papillae. *Biochimica et Biophysica Acta, 208*(2), 219–226.

Fraser, J. R., Alcorn, D., Laurent, T. C., Robinson, A. D., & Ryan, G. B. (1985). Uptake of circulating hyaluronic acid by the rat liver. *Cell and Tissue Research, 242*(3), 505–510.

Fraser, J. R., Kimpton, W. G., Pierscionek, B. K., & Cahill, R. N. (1993). The kinetics of hyaluronan in normal and acutely inflamed synovial joints: Observations with experimental arthritis in sheep. *Seminars in Arthritis and Rheumatism, 6*(1), 9–17.

Giji, S., Arumugam, M., Abirami, P., & Balasubramanian, T. (2012). Isolation and characterization of hyaluronic acid from the liver of marine stingray *Aetobatus narinari*. *International Journal of Biological Macromolecules, 54*, 84–89.

Hadidian, Z., & Pirie, N. W. (1948). The preparation and some properties of hyaluronic acid from human umbilical cord. *Biochemical Journal, 42*(2), 260–265.

Hakim, A., & Linhardt, R. J. (1990). Isolation and recovery of acidic oligosaccharides from polyacrylamide gels by semi-dry electrotransfer. *Electrophoresis, 11*, 23–28.

Hallgren, R., Gerdin, B., & Tufveson, G. (1990). Hyaluronic acid accumulation and redistribution in rejecting rat kidney graft. *Journal of Experimental Medicine, 171*, 2063–2076.

Holmbeck, S., & Lerner, L. (1993). Separation of hyaluronan oligosaccharides by the use of anion-exchange HPLC. *Carbohydrate Research, 239*, 239–244.

Homer, K., Denbow, L., Whiley, R., & Beighton, D. (1993). Chondroitin sulfate depolymerase and hyaluronidase activities of viridians streptococci determined by a sensitive spectrophotometric assay. *Journal of Clinical Microbiology, 31*(6), 1648–1651.

Kanchana, S., Arumugam, M., Giji, S., & Balasubramanian, T. (2013). Isolation, characterization and antioxidant activity of hyaluronic acid from marine bivalve mollusk *Amussium pleuronectus* (Linnaeus, 1758). *Bioactive Carbohydrates and Dietary Fibre, 2,* 1–7.

Kang, D. Y., Kim, W. S., Heo, I. S., Park, Y. H., & Lee, S. (2010). Extraction of hyaluronic acid (HA) from rooster comb and characterization using flow field-flow fractionation (FlFFF) coupled with multiangle light scattering (MALS). *Journal of Separation Science, 33*(22), 3530–3536.

Kay, R. E., Walwick, E. R., & Gilford, C. K. (1954). Spectral change in a cationic dye due to interaction with macromolecules. II. Effects of environment and macromolecule structure. *Journal of Physical Chemistry, 68,* 1907–1916.

Kim, Y. S., Jo, Y. Y., Chang, I. M., Toida, T., Park, Y., & Linhardt, R. J. (1996). A new glycosaminoglycan from the giant African snail *Achatina fulica. Journal of Biological Chemistry, 271,* 11750–11755.

Kuhlman, R. E., & Kaufman, H. E. (1960). A microchemical study of the aqueous humor enzyme-protein interrelations. *A.M.A. Archives of Ophthalmology, 63*(1), 41–46.

Laurent, U. B., & Fraser, J. R. (1983). Turnover of hyaluronate in the aqueous humour and vitreous body of the rabbit. *Experimental Eye Research, 36*(4), 493–503.

Laurent, T. C., & Fraser, J. R. (1992). Hyaluronan. *FASEB Journal, 6,* 2397–2404.

Lebel, L., Smith, L., Risberg, B., Laurent, T. C., & Gerdin, B. (1988). Effect of increased hydrostatic pressure on lymphatic elimination of hyaluronan from sheep lung. *Journal of Applied Physiology, 64,* 1327–1332.

Lee, J. Y., & Spicer, A. P. (2000). Hyaluronan: A multifunctional, megadalton, stealth molecule. *Current Opinion in Cell Biology, 12,* 581–586.

Lesley, J., Hascall, V. C., Tammi, M., & Hyman, R. (2000). Hyaluronan binding by cell surface CD44. *Journal of Biological Chemistry, 275,* 26967–26975.

Li, J., Kisara, K., Danielsson, S., & Lindstrom, M. (2007). An improved methodology for the quantification of uronic acid units in xylans and other polysaccharides. *Carbohydrate Research, 342*(11), 1442–1449.

Liu, L., Liu, Y., Li, J., Du, G., & Chen, J. (2011). Microbial production of hyaluronic acid: Current state, challenges, and perspectives. *Microbial Cell Factories, 10,* 99.

Maccari, F., & Volpi, N. (2003). Detection of submicrogram quantities of Escherichia coli lipopolysaccharides by agarose-gel electrophoresis. *Analytical Biochemistry, 322*(2), 185–189.

Min, H., & Cowman, M. K. (1986). Combined alcian blue and silver staining of glycosaminoglycans in polyacrylamide gels: Application to electrophoretic analysis of molecular weight distribution. *Analytical Biochemistry, 155*(2), 275–285.

Murado, M. A., Montemayor, M. I., Cabo, M. L., Vazquez, J. A., & Gonzalez, M. P. (2012). Optimization of extraction and purification process of hyaluronic acid from fish eyeball. *Food and Bioproducts Processing, 90*(3), 491–498.

Nader, H. B., Ferreira, T. M., Paiva, J. F., Medeiros, M. G., Jeronimo, S. M., Paiva, V. M., et al. (1984). Isolation and structural studies of heparin sulphates and chondroitin sulphates from three species of molluscs. *Journal of Biological Chemistry, 259,* 1431–1435.

Necas, J., Bartosikova, L., Brauner, P., & Kolar, J. (2008). Hyaluronic acid (hyaluronan): A review. *Veterinární Medicína, 53*(8), 397–411.

Nettelbladt, O., Bergh, J., Schenholm, M., Tengblad, A., & Hallgren, R. (1989). Accumulation of hyaluronic acid in the alveolar interstitial tissue in bleomycin-induced alveolitis. *American Review of Respiratory Disease, 139*(3), 759–762.

Nishikawa, S., & Tamai, M. (1996). Ultrastructure of hyaluronic acid and collagen in the human vitreous. *Current Eye Research, 15*(1), 37–43.

Ogston, A. G., & Sherman, T. F. (1958). Electrophoretic removal of protein from hyaluronic acid. *Nature, 181*(4607), 482–483.

Oliveira, F. W., Chavante, S. F., Santos, E. A., Dietrich, C. P., & Nader, H. B. (1994). Appearance and fate of a beta-galactanase, alpha, beta-galactosidases, heparan sulphate and chondroitin sulphate degrading enzymes during embryonic development of the mollusc *Pomacea* sp. *Biochimica et Biophysica Acta, 1200,* 241–246.

Orvisky, E., Soltes, L., Chabrcek, P., Novak, I., Kery, V., Stancikova, M., et al. (1992). The determination of hyaluronan molecular weight distribution by means of high-performance size exclusion chromatography. *Journal of Liquid Chromatography, 15*(18), 3203–3218.

Pethrick, R. A., Ballada, A., & Zaikov, G. E. (2007). Handbook of polymer research: Monomers, oligomers, polymers and composites. In Antonio Ballada, & Gennadiĭ Efremovich Zaikov (Eds.), New York: Nova Science Publishers, Inc.

Postlethwaite, A. E., Smith, G. N., Jr., Lachman, L. B., Endres, R. O., Poppleton, H. M., Hasty, K. A., et al. (1989). Stimulation of glycosaminoglycan synthesis in cultured human dermal fibroblasts by interleukin 1. Induction of hyaluronic acid synthesis by natural and recombinant interleukin 1s and synthetic interleukin 1 beta peptide 163-171. *Journal of Clinical Investigation, 83*(2), 629–636.

Price, R. D., Berry, M. G., & Navsaria, H. A. (2007). Hyaluronic acid: The scientific and clinical evidence. *Journal of Plastic, Reconstructive & Aesthetic Surgery, 60,* 1110–1119.

Pryce-Jones, R. H., & Lannigan, N. A. (1979). Hyaluronidase: A colourimetric assay. *Journal of Pharmacy and Pharmacology, 31,* 92p.

Reissig, J. L., Strominger, J. L., & Leloir, L. F. (1955). A modified colorimetric method for the estimation of N-acetylamino sugars. *Journal of Biological Chemistry, 217,* 959–966.

Rice, K. G., Rottink, M. K., & Linhardt, R. J. (1987). Fractionation of heparin-derived oligosaccharides by gradient polyacrylamide-gel electrophoresis. *Biochemical Journal, 244,* 515–522.

Saranraj, P., & Naidu, M. A. (2013). Hyaluronic acid production and its applications—A review. *International Journal of Pharmaceutical and Biological Archive, 4*(5), 853–859.

Schmidt, M. (1962). Fractionation of acid mucopolysaccharides on DEAE-Sephadex anion exchanger. *Biochimica et Biophysica Acta, 63,* 346–348.

Scott, J. E. (1960). Aliphatic ammonium salts in the assay of acidic polysaccharides from tissues. *Methods of Biochemical Analysis, 8,* 145–197.

Seastone, C. V. (1939). The virulence of group C haemolytic streptococci of animal origin. *The Journal of Experimental Medicine, 70,* 361–378.

Senni, K., Pereira, J., Gueniche, F., Delbarre-Ladrat, C., Sinquin, C., Ratiskol, J., et al. (2011). Marine polysaccharides: A source of bioactive molecules for cell therapy and tissue engineering. *Marine Drugs, 9,* 1664–1681.

Shinomiya, K., Kabasawa, Y., Toida, T., Imanari, T., & Ito, Y. (2001). Separation of chondroitin sulphate and hyaluronic acid fragments by centrifugal precipitation chromatography. *Journal of Chromatography A, 922,* 365–369.

Sven, B., & Zhag, T. (1986). *Yiyao Gogye, 17,* 291.

Takagaki, K., Nakamura, T., Majima, M., & Endo, M. (1988). Isolation and characterization of a chondroitin sulfate-degrading endo-beta-glucuronidase from rabbit liver. *Journal of Biological Chemistry, 263,* 7000–7006.

Tammi, R., MacCallum, D., Hascall, V. C., Pienimaki, J. P., Hyttinen, M., & Tammi, M. (1998). Hyaluronan bound to CD44 on keratinocytes is displaced by hyaluronandecasaccharides and not hexasaccharides. *Journal of Biological Chemistry, 273,* 28878–28888.

Toffanin, R., Kvam, B. J., Flaibani, A., Atzori, M., Biviano, F., & Paoletti, S. (1993). NMR studies of oligosaccharides derived from hyaluronate: Complete assignment of ^{1}H and ^{13}C NMR spectra of aqueous di- and tetra-saccharides and comparison of chemical shifts for oligosaccharides of increasing degree of polymerisation. *Carbohydrate Research, 245*, 113–128.

Tommeraas, K., & Melander, C. (2008). Kinetics of hyaluronan hydrolysis in acidic solution at various pH values. *Biomacromolecules, 9*, 1535–1540.

Toole, B. P. (2004). Hyaluronan: From extracellular glue to pericellular cue. *Nature Reviews. Cancer, 4*, 528–539.

Toyoda, H., Motoki, K., Tanikawa, M., Shinomiya, K., Akiyama, H., & Imanari, T. (1991). Determination of human urinary hyaluronic acid, chondroitin sulphate and dermatan sulphate as their unsaturated disaccharides by high-performance liquid chromatography. *Journal of Chromatography, 565*(1–2), 141–148.

Vazquez, J. A., Montemayor, M. I., Fraguas, J., & Murado, M. A. (2009). High production of hyaluronic and lactic acids by *Streptococcus zooepidemicus* in fed-batch cultures using commercial and marine peptones from fishing by-products. *Biochemical Engineering Journal, 44*, 125–130.

Vazquez, J. A., Montemayor, M. I., Fraguas, J., & Murado, M. A. (2010). Hyaluronic acid production by Streptococcus zooepidemicus in marine by-products media from mussel processing wastewaters and tuna peptone viscera. *Microbial Cell Factories, 9*, 46–56.

Vieira, R. P., & Mourao, P. A. (1988). Occurrence of a unique fucose-branched chondroitin sulphate in the body wall of a sea cucumber. *Journal of Biological Chemistry, 263*, 18176–18183.

Volpi, N. (1996). Electrophoresis separation of glycosaminoglycans on nitrocellulose membranes. *Analytical Biochemistry, 240*(1), 114–118.

Volpi, N. (2003). Milligram-scale preparation and purification of oligosaccharides of defined length possessing the structure of chondroitin from defructosylated capsular polysaccharide K4. *Glycobiology, 13*(9), 635–640.

Volpi, N. (2004a). Separation of capsular polysaccharide K4 and defructosylated K4 by high-performance capillary electrophoresis. *Electrophoresis, 25*(4–5), 692–696.

Volpi, N. (2004b). Purification of the Escherichia coli K5 capsular polysaccharide and use of high-performance capillary electrophoresis to qualitative and quantitative monitor the process. *Electrophoresis, 25*(18–19), 3307–3312.

Volpi, N., & Maccari, F. (2005). Microdetermination of chondroitin sulfate in normal human plasma by fluorophore-assisted carbohydrate electrophoresis (FACE). *Clinica Chimica Acta, 356*(1–2), 125–133.

Volpi, N., & Maccari, F. (2006). Electrophoretic approaches to the analysis of complex polysaccharides. *Journal of Chromatography B, 834*, 1–13.

Wessler, E. (1968). Analytical and preparative separation of acidic glycosaminoglycans by electrophoresis in barium acetate. *Analytical Biochemistry, 26*(3), 439–444.

Yamada, S., Sugahara, K., & Ozbek, S. (2011). Evolution of glycosaminoglycans. *Communicative & Integrative Biology, 4*(2), 150–158.

Yeung, B., & Marecak, D. (1999). Molecular weight determination of hyaluronic acid by gel filtration chromatography coupled to matrix-assisted laser desorption ionization mass spectrometry. *Journal of Chromatography. A, 852*(2), 573–581.

Zhang, W., Moskowitz, R. W., Nuki, G., Abramson, S., Altman, R. D., Arden, N., et al. (2007). OARSI recommendations for the management of hip and knee osteoarthritis. Part II: OARSI evidence-based, expert consensus guidelines. *Osteoarthritis and Cartilage, 16*(2), 137–162.

CHAPTER FIVE

Extracellular Polysaccharides Produced by Marine Bacteria

Panchanathan Manivasagan*, Se-Kwon Kim[†,1]

*Marine Biotechnology Laboratory, Department of Chemistry, Pukyong National University, Busan, South Korea
[†]Department of Marine-bio Convergence Science, Specialized Graduate School Science and Technology Convergence, Marine Bioprocess Research Center, Pukyong National University, Busan, South Korea
[1]Corresponding author: e-mail address: sknkim@pknu.ac.kr

Contents

1. Introduction	80
2. Extracellular Polysaccharides	81
3. Roles of Microbial EPS in the Marine Environment	81
4. Biosynthesis	83
5. Source of Extracellular Polysaccharide-Producing Bacteria	83
6. Isolation of Extracellular Polysaccharide-Producing Bacteria	83
7. Marine EPS-Producing Microorganisms	83
7.1 Marine bacteria	83
7.2 Marine cyanobacteria	87
7.3 Marine actinobacteria	88
8. Biotechnological Applications of Extracellular Polysaccharides	89
8.1 Medicinal applications	89
8.2 Gelling agent	89
8.3 Emulsifiers	90
8.4 Heavy metal removal	90
8.5 Enhanced oil recovery	90
9. Conclusions	90
Acknowledgments	91
References	91

Abstract

Extracellular polysaccharides (EPSs) produced by microorganisms are a complex mixture of biopolymers primarily consisting of polysaccharides, as well as proteins, nucleic acids, lipids, and humic substances. Microbial polysaccharides are multifunctional and can be divided into intracellular polysaccharides, structural polysaccharides, and extracellular polysaccharides or exopolysaccharides. Recent advances in biological techniques allow high levels of polysaccharides of interest to be produced *in vitro*. Biotechnology is a powerful tool to obtain polysaccharides from a variety of marine microorganisms, by controlling the growth conditions in a bioreactor while tailoring the production of biologically active compounds. The aim of this chapter is to give an overview of current knowledge on extracellular polysaccharides producing marine bacteria isolated from marine environment.

1. INTRODUCTION

Polysaccharides occur as important constituents of plant and microbial cell walls, either as storage polysaccharides or as biopolymers known as exopolysaccharides (EPSs) secreted by microorganisms. In recent years, there has been a growing interest in the isolation and identification of new microbial polysaccharides that might have novel uses such as viscosifiers, gelling agents, emulsifiers, stabilizers, and texture enhancers (Guezennec, 2002).

The increased demand for natural polymers for various industrial applications in recent years has led to a renewed interest in exopolysaccharide production by microorganisms. Many microorganisms have an ability to synthesize extracellular polysaccharides and excrete them out of cell either as soluble or as insoluble polymers (Suresh Kumar, Mody, & Jha, 2007). Various exopolysaccharides produced by bacteria have novel and unique physical characteristics and are generally referred to water-soluble gums (Morin, 1998). Exopolysaccharides have found extensive applications in food, pharmaceutical, and other industries. Many species of gram-positive and gram-negative bacteria, fungi, and also some algae are known to produce exopolysaccharides. Considering the biodiversity of the microbial world and the number of articles published each year on new microbial exopolysaccharides, it is astonishing to realize that only three of them (i.e., dextran, xanthan, and gellan gums) have survived the industrial competition (Suresh Kumar et al., 2007).

For several years, there has been a growing interest in the recognition of biological activities of microbial polysaccharides (Weiner, Langille, & Quintero, 1995) such as their antitumor activity and the immunostimulatory activities of some polysaccharides produced by marine bacteria, for example, *Vibrio* sp. and *Pseudomonas* sp. (Matsuda, Shigeta, & Okutani, 1999; Okutani, 1985). Antiviral effects of polysaccharides including human immunodeficiency virus activity have been recognized for a decade (Baba et al., 1998; Tzianabos, 2000). The anticoagulant activity of these polymers can also be linked to the high sulfate content associated with specific polysaccharides (Nishino & Nagumo, 1992; Volpi, Sandri, & Venturelli, 1995).

A number of microbial extracellular polysaccharides are produced on an industrial scale. In addition to dextran and xanthan, gellan gum is gaining importance in the food industry (Kano & Pettitt, 1993). Other well-known polysaccharides of microbial origin include curdlan secreted by *Alcaligenes faecalis* var *myxogenes*; alginates from *Pseudomonas aeruginosa* and *Azotobacter vinelandii*; succinoglycans produced by bacteria of the genera *Pseudomonas, Rhizobium, Agrobacterium*, and *Alcaligenes*; and others (Sutherland, 1996). The aim of this chapter is to give an overview of current knowledge on extracellular polysaccharides producing marine bacteria isolated from marine environment.

2. EXTRACELLULAR POLYSACCHARIDES

Microbial polysaccharides represent a class of important products that are of growing interest for many sectors of industry. The advantages of microbial polysaccharides over plants polysaccharides are their novel functions and constant chemical and physical properties. A number of common marine bacteria widely distributed in the oceans can produce EPSs; nevertheless, most of these EPSs remain poorly understood, and only a few of them have been fully characterized. In recent years, there has been a growing interest in isolating new exopolysaccharides (EPSs)-producing bacteria from marine environments, particularly from various extreme marine environments (Nichols, Guezennec, & Bowman, 2005). Many new marine microbial EPSs with novel chemical compositions, properties, and structures have been found to have potential applications in fields such as adhesives, textiles, pharmaceuticals, and medicine for anticancer, food additives, oil recovery, and metal removal in mining and industrial waste treatments, etc. General information about the EPSs produced by marine bacteria, including their chemical compositions, properties, and structures, together with their potential applications in industry, are widely reported (Weiner et al., 1995).

3. ROLES OF MICROBIAL EPS IN THE MARINE ENVIRONMENT

The physiological role of EPS depends on the ecological niches and the natural environment in which microorganisms have been isolated as well. Indeed, the EPS production is a process that requires a noticeable energy cost of up to 70%, representing a significant carbon investment for microorganisms. However, the benefits related to EPSs are significantly higher than costs considering the increasing growth and survival of microorganisms in their presence (Wolfaardt, Lawrence, & Korber, 1999). Indubitably, they possess a protective nature: the EPSs, forming a layer surrounding a cell, provide an effective protection against high or low temperature and salinity or against possible predators. They are essential in the aggregate formation, in the mechanism of adhesion to surfaces and to other organisms, in the formation of biofilm, and in the uptake of nutrients (Alldredge, 2000; Decho & Herndl, 1995; Holmström & Kjelleberg, 1999; Sutherland, 2001). In particular, studies of sea ice microbial communities have also found bacteria strongly associated to particles and have pointed out that microbial EPS played an important role in cryoprotection (Krembs, Eicken, Junge, & Deming, 2002; Logan & Hunt, 1987). Moreover, the rate of synthesis and the amount of exopolysaccharide accumulated as capsular form in pathogenic bacteria influence their pathogenicity, in that capsular polysaccharides do not stimulate the immune system for their ability to mimic the cell surface of host cells.

EPSs display an important role in biofilm matrix in regard to the biochemical interactions between bacteria and surrounding cells (Logan & Hunt, 1987). The hydrated biofilms offer a stable microenvironment in which extracellular enzymes can find storage and in the same time facilitating cellular uptake of small molecules (Decho & Herndl, 1995). In addition, in a natural aquatic environment, the nutrients can interact with exopolymers in order to increase the rate of substance uptake and concentrate the dissolved organic compounds, making them available to support microbial growth (Decho & Herndl, 1995). Finally, it has been proved that strains isolated from deep-sea hydrothermal vents show resistance to heavy metals and their purified EPSs presented the capacity to bind metals and toxic substances (Loaëc, Olier, & Guezennec, 1997, 1998; Wuertz, Müller, Spaeth, Pfleiderer, & Flemming, 2000). Indeed, these exopolymers exhibit a polyanionic state in marine environment displaying a high-binding affinity for cations as well as trace metals. Since they generally contain uronic acids with a percentage between 20% and 50% of the total polysaccharide content, the acidic carboxyl groups attached are ionizable at seawater pH. Furthermore, EPS produced by some Antarctic bacterial isolates contain uronic acids and sulfate groups and may act as ligands for cations present as trace metals in the Southern Ocean environment, enhancing the primary production of microbial communities usually limited by poor availability of trace metals such as iron (Fe^{3+}) (Nichols, Lardière, et al., 2005).

4. BIOSYNTHESIS

Extensive progress has been made in recent years in determining the biosynthetic and genetic mechanisms involved in the synthesis of exopolysaccharides. The biosynthesis of most exopolysaccharides closely resembles the process by which the bacterial cell wall polymer, peptidoglycan, and lipopolysaccharide are formed. Indeed, the three types of macromolecules share the characteristic of being synthesized at the cell membrane and being exported to final site external to the cytoplasmic membrane. The only exceptions are levans, alternans, and dextrans, which are synthesized by an extracellular process (Vanhooren & Vandamme, 1998).

5. SOURCE OF EXTRACELLULAR POLYSACCHARIDE-PRODUCING BACTERIA

Microorganisms producing extracellular polysaccharide are found in various ecological niches. Environment offering a medium with high-carbon/nitrogen ratio is known to contain microorganisms producing polysaccharides, for example, effluents from the sugar, paper, or food industries as well as wastewater plants (Morin, 1998).

6. ISOLATION OF EXTRACELLULAR POLYSACCHARIDE-PRODUCING BACTERIA

Extracellular polysaccharide-producing organisms can be isolated using a complex media or a chemically defined synthetic media. These organisms produce colonies with mucoid or watery surface and thus can be detected macroscopically. Morin stated that no direct correlation exists between morphological characteristics of colonies on solid medium and the ability of a culture to produce polysaccharides in liquid medium. Some polysaccharides might form stable complexes with water-soluble dyes such as aniline blue, which could be used as a screening tool (Morin, 1998).

7. MARINE EPS-PRODUCING MICROORGANISMS

7.1. Marine bacteria

Exopolysaccharides (EPSs) may have an important role in the Antarctic marine environment, possibly acting as ligands for trace metal nutrients such as iron or providing cryoprotection for growth at low temperature and high salinity (Nichols, Lardière, et al., 2005). Mancuso Nichols et al. reported the production of exopolysaccharides by Antarctic marine bacteria isolated from particulate material sampled from seawater and sea ice in the southern ocean. Analyses of 16S rDNA sequences placed these isolates in the genus *Pseudoalteromonas*. In batch culture, strain produced EPS. The yield of EPS produced by *Pseudoalteromonas* was 30-fold higher at −2 and 10 °C than at 20 °C. Crude chemical analyses showed that these EPSs were composed primarily of neutral sugars and uronic acids with sulfates. Gas chromatographic analysis of monosaccharides confirmed these gross compositional findings and molar ratios of monosaccharides revealed differences between the two EPSs (Mancuso Nichols, Garon, Bowman, Raguénès, & Guezennec, 2004).

Nichols et al. reported the chemical characterization of exopolysaccharides from Antarctic marine bacteria isolated from Southern Ocean particulate material or from sea ice. Whole cell fatty acid profiles and 16S rRNA gene sequences showed that the isolates included representatives of the genera *Pseudoalteromonas*, *Shewanella*, *Polaribacter*, and *Flavobacterium* as well as one strain, which constituted a new bacterial genus in the family Flavobacteriaceae. The isolates are, therefore, members of the "*Gammaproteobacteria*" and *Cytophaga-Flexibacter-Bacteroides*, the taxonomic groups that have been shown to dominate polar sea ice and seawater microbial communities. Exopolysaccharides produced by Antarctic isolates were characterized. Chemical composition and molecular weight data revealed that these EPSs were very diverse, even among six closely related *Pseudoalteromonas* isolates. Most of the EPSs contained charged uronic acid residues; several also contained sulfate groups. Some strain produced unusually large polymers (molecular weight up to 5.7 MDa) including one strain in which EPS synthesis is stimulated by low temperature (Nichols, Lardière, et al., 2005).

Christensen et al. reported the partial chemical and physical characterization of two extracellular polysaccharides produced by marine, periphytic *Pseudomonas* sp. strain NCMB 2021, which can attach to solid, and especially hydrophobic, surfaces, and elaborates two different extracellular polysaccharides in batch cultures. One (polysaccharide A) was produced only during exponential growth and contained glucose, galactose, glucuronic acid, and galacturonic acid in a molar ratio of 1.00:0.81:0.42:0.32. It produced viscous solutions, formed gels at high concentrations, and precipitated with several multivalent cations. The other (polysaccharide B) was released at the end of the exponential phase and in the stationary phase. It contained equimolar amounts of N-acetylglucosamine, 2-keto-3-deoxyoctulosonic acid, an unidentified 6-deoxyhexose, and also O-acetyl groups (Christensen, Kjosbakken, & Smidsrød, 1985). Boyle et al. reported the characterization of two extracellular polysaccharides from marine bacteria isolated from the intertidal zone about 30 km east of Halifax, Nova Scotia (Boyle & Reade, 1983).

Iwabuchi et al. reported that the extracellular polysaccharides of *Rhodococcus rhodochrous* S-2 stimulate the degradation of aromatic components in crude oil by indigenous marine bacteria. *R. rhodochrous* S-2 produces extracellular polysaccharides (S-2 EPSs) containing D-glucose, D-galactose, D-mannose, D-glucuronic acid, and lipids, which are important to the tolerance of this strain to an aromatic fraction (AF) of Arabian light crude oil. S-2 EPS was, hence, the most effective of the surfactants tested in promoting the biodegradation of AF and may thus be an attractive agent to use in the bioremediation of oil-contaminated marine environments (Iwabuchi et al., 2002).

Kumar et al. reported the purification and characterization of an extracellular polysaccharide from haloalkalophilic *Bacillus* sp. I-450 isolated from soil samples collected from different zones of the heavily polluted tidal mudflats of the Korean Yellow Sea around Inchon City. The isolate produced the polysaccharide during the late logarithmic growth phase. The polymer could be recovered from the supernatant of the fermented medium by cold ethanol precipitation and purified by treating with cetylpyridinium chloride. The polymer was identified as an acidic polysaccharide containing neutral sugars, namely, galactose, fructose, glucose, and raffinose, and uronic acids as major and minor components, respectively. The amount of neutral sugars, uronic acid, and amino sugars was 52.4%, 17.2%, and 2.4%, respectively. The molecular weight of the polysaccharide was found to be 2.2×10^6 Da (Kumar, Joo, Choi, Koo, & Chang, 2004). Lee et al. reported the extracellular polysaccharide-producing marine bacterium, *Hahella chejuensis* gen. nov., sp. nov., isolated from marine sediment collected from Marado, Cheju Island, Republic of Korea (Lee et al., 2001).

Iyer et al. reported the exopolysaccharide producing *Enterobacter cloaceae* (AK-I-MB-71a) isolated from a marine sediment sample from the west coast of India and was tested for its Cr(VI) tolerance. This isolate was not only resistant to this heavy metal but also showed enhanced growth and exopolysaccharide production in the presence of Cr(VI) at 25, 50, and 100 ppm concentrations (Iyer, Mody, & Jha, 2004).

Raguenes et al. reported the new polysaccharide producing bacterium, *Alteromonas infernus* sp. nov., isolated from a deep-sea hydrothermal vent. During the stationary phase in batch cultures in the presence of glucose, this bacterium secreted two unusual polysaccharides. The water-soluble exopolysaccharide-1 produced contained glucose, galactose, galacturonic, and glucuronic acids as monosaccharides. The gel-forming exopolysaccharide-2 was separated from the bacterial cells by dialysis against distilled water and partially characterized (Raguénès et al., 1997).

Two water-soluble extracellular polysaccharides, ETW1 and ETW2, were isolated from the marine bacterium *Edwardsiella tarda* by ion-exchange and size-exclusion chromatography, and their structures were investigated. ETW1 and ETW2 are mannans, with molecular weights of about 29 and 70 kDa, respectively. Antioxidant properties of the two extracellular polysaccharides were evaluated with hydroxyl and 1,1-diphenyl-2-picrylhydrazyl (DPPH) radicals scavenging activities and lipid peroxidation inhibition *in vitro*, and the results showed that ETW1 and ETW2 had good antioxidant and hydroxyl and DPPH radicals scavenging activities. ETW1 exhibited higher antioxidant activity than ETW2 and could be a potential source of antioxidant and used as possible food supplement or ingredient in the pharmaceutical industry (Guo et al., 2010).

Fang et al. reported the production of antioxidant extracellular polysaccharides from *Bacillus licheniformis*, isolated from marine mud. The maximum yield of the extracellular polysaccharides of *B. licheniformis* OSTK95 was 68.59 mg L^{-1} after treated by 10% *n*-hexane or 1% xylene for 3 h, while the maximum yield of the extracellular polysaccharides of strain UD061 was 185.01 mg L^{-1} after treated by 12.5% *n*-hexane or 5% xylene for 3 h. Finally, the continuous passage experiment showed that the strains have high genetic stability (Fang, Liu, Lu, Jiao, & Wang, 2012). Fang et al. reported the optimization of physical and nutritional variables for the production of antioxidant exopolysaccharides by *B. licheniformis* UD061 in solid-state fermentation with squid-processing by-product and maize cob meal used as carbon and nitrogen source and solid matrix. Maximum EPS production was obtained under the optimal conditions of 4.08 g L^{-1} NaCl, 0.71 g L^{-1} $MgSO_4 \cdot 7H_2O$, and 60.40% moisture level. A production of 1468 mg gds^{-1}, which was well in agreement with the predicted value, was achieved using this optimized procedure (Fang et al., 2013).

Antarctic bacteria are a novel source of polysaccharides, which might have potential applications as biological response modifiers. A heteropolysaccharide partially esterified polysaccharide (PEP) was isolated from the liquid culture broth of the Antarctic bacterium *Pseudoaltermonas* sp. S-5. PEP contained mannose, glucose, and galactose in a ratio of 4.8:50.9:44.3. High-performance gel permeation chromatography of this polysaccharide showed a unimodal profile, and the molecular weight was 397 kDa. PEP was studied for its immunological effects on peritoneal macrophage cells (Bai et al., 2012).

7.2. Marine cyanobacteria

Cyanobacteria, also known as blue-green algae or blue-green bacteria because of their color (the name comes from Greek: kyanós = blue), are a significant component of the marine nitrogen cycle and an important primary producer in many areas of the ocean. They are also found in habitats other than the marine environment; in particular, cyanobacteria are known to occur in both freshwater, hypersaline inland lakes, and arid areas where they are a major component of biological soil crusts (Laurienzo, 2010).

Since the early 1950s, more than 100 cyanobacteria strains of different genera have been investigated regard to the production of exocellular polysaccharides. Such polysaccharides are present as outermost investments forming sheaths, capsules, and slimes that protect the bacterial cells from the environment. Moreover, most polysaccharide-producing cyanobacteria release aliquots of capsules and slimes as soluble polymers in the culture medium (De Philippis, Sili, Paperi, & Vincenzini, 2001). In recent years, the interest toward such cyanobacteria has greatly increased, in particular toward those strains that possess abundant capsules and slimes and so release large amount of soluble polysaccharides, which can be easily recovered from liquid culture.

Cyanobacteria, which can be a potential source of polysaccharide, are yet unexploited (De Philippis, Margheri, Materassi, & Vincenzini, 1998). A marine nitrogen-fixing cyanobacterium, *Cyanothece* sp. ATCC 51142, has an ability to produce extracellular polysaccharide (EPS) at a high level (Shah, Ray, Garg, & Madamwar, 2000). Shah et al. reported the EPS production by a marine cyanobacterium *Cyanothece* sp. EPS production was found to be influenced by the concentration of salt, pH, and type of nitrogen source. Maximum polysaccharide production was found to occur at a 4.5% (w/v) NaCl salt concentration, pH 7.0, and in the presence of $NaNO_3$ as the nitrogen source. The gelation of EPS in alkaline conditions was employed to remove the dyes from the effluents. The effect of organic molecules and metal ions on the efficiency of dye removal capacity was investigated. A laboratory-scale reactor was prepared to treat artificial textile effluent (Shah, Garg, & Madamwar, 1999).

Shah et al. reported the characterization of the extracellular polysaccharide produced by a marine cyanobacterium, *Cyanothece* sp. ATCC 51142. Physical analysis of the exopolysaccharide (EPS), such as nuclear magnetic resonance and infrared spectrum, was done to determine its possible structure. Thermal gravimetric analysis, differential scanning calorimeter, and differential thermal analysis of the polymer were done to find out the thermal behavior. Calcium content within the sample was found out. Some of the physicochemical properties, such as relative viscosity, specific viscosity, and intrinsic viscosity of the EPS, were studied under different conditions. The phenomenon of gel formation by the EPS was investigated for its potential application in metal removal from solutions. De Philippis et al. reported the potential of unicellular cyanobacteria; *Cyanothece* strains isolated from saline environments have been characterized with regard to exopolysaccharide production (De Philippis et al., 1998).

Chi et al. reported the new exopolysaccharide produced by marine cyanobacterium, *Cyanothece* sp. 113 isolated from the sea in China. *Cyanothece* sp. 113, a unicellular, aerobic, diazotrophic and photosynthetic marine cyanobacterium, produced 22.34 g/l or exopolysaccharide in 11 days at 29 °C, aeration rate of 7.0 l/min and continuous illumination with 4300 lux (Chi, Su, & Lu, 2007). *Cyanothece* sp. 113 is a unicellular, aerobic, diazotrophic, and photosynthetic marine cyanobacterium. The optimal medium for exopolysaccharide yield by the strain was 70.0 g L^{-1} of NaCl and 0.9 g L^{-1} of $MgSO_4$ based on the modified F/2 medium for cultivation of marine algae. The optimal cultivation condition for exopolysaccharide yield by this cyanobacterial strain was 29 °C, aeration, and continuous illumination at 86.0 $\mu E\ M^{-2}\ S^{-1}$. Under the optimal conditions, over 18.4 g L^{-1} of exopolysaccharide was produced within 12 days (Su, Chi, & Lu, 2007).

7.3. Marine actinobacteria

Marine actinobacteria are one of the most efficient groups of secondary metabolite producers and are very important from an industrial point of view. Many representatives of the order *Actinomycetales* are prolific producers of thousands of biologically active metabolites. A variety of extracellular polysaccharides have been isolated from bacteria and fungi (Barreto-Bergter & Gorin, 1983; Jansson, Kenne, & Lindberg, 1975). Some of them are utilized commercially. Polysaccharide produced by *Streptomyces* species, however, has not been fully studied. Inoue et al. reported the production and partial properties of an extracellular polysaccharide from *Streptomyces* sp. A-1845 (Inoue, Murakawa, & Endo, 1992). Manivasagan et al. reported the production and characterization of an extracellular polysaccharide from *Streptomyces violaceus* MM72, isolated from the marine sediments of Tuticorin coast, India. Medium composition and culture conditions for the EPS production by *S. violaceus* MM72 were optimized using two statistical methods: Plackett–Burman design applied to find the key ingredients and conditions for the best yield of EPS production and central composite design used to optimize the concentration of the three significant variables: glucose, tryptone, and NaCl. The results showed the *S. violaceus* MM72 produced a kind of EPS having molecular weight of 8.96×10^5 Da (Manivasagan et al., 2013).

8. BIOTECHNOLOGICAL APPLICATIONS OF EXTRACELLULAR POLYSACCHARIDES

Extracellular polysaccharides find their applications in various fields ranging from medicinal applications to be used as a source of monosaccharides.

8.1. Medicinal applications

Antitumor, antiviral, and immunostimulant activities of polysaccharides produced by marine *Vibrio* and *Pseudomonas* have been reported by Okutani (1984, 1992). A low-molecular-weight heparin-like exopolysaccharide exhibiting anticoagulant property has been isolated from *A. infernus*, obtained from deep-sea hydrothermal vents (Senni et al., 2011). Clavan, an L-fucose containing polysaccharide, has a potential application in preventing tumor cell colonization of the lung, in controlling the formation of white blood cells, in the treatment of the rheumatoid arthritis, in the synthesis of antigens for antibody production, and in cosmeceuticals as skin-moisturizing agent (Vanhooren & Vandamme, 2000).

8.2. Gelling agent

Gelrite, obtained from *Pseudomonas* spp., is a new gelling polysaccharide with good thermal stability and clarity. It has been reported that gelrite is superior to agar (Lin & Casida, 1984). It forms a brittle, firm, and optically clear gel upon deacetylation using mild alkali (Kang, Veeder, Mirrasoul, Kaneko, & Cottrell, 1982).

8.3. Emulsifiers

Surfactants and emulsifiers from bacterial sources have attracted attention because of their biodegradability and possible production from renewable resources. Emulsan produced by *Acinetobacter calcoaceticus* RAG-1 has been commercialized (Rosenberg, Zuckerberg, Rubinovitz, & Gutnick, 1979). A viscous exopolysaccharide has also been isolated from *Sphingomonas paucimobilis*. This polysaccharide stabilized emulsions more effectively than other commercial gums such as arabic, tragacanth, karaya, and xanthan (Ashtaputre & Shah, 1995). Apart for Emulsan, an exopolysaccharide produced by a marine bacteria, is reported to form stable emulsions with a number of hydrocarbons. This exopolysaccharide proved to be more efficient than the commercially available emulsifiers (Iyer, Mody, & Jha, 2006).

8.4. Heavy metal removal

Contamination of the environment by heavy metals is of growing concern because of the health risks posed to humanity and animals. Cell-bound polysaccharide produced by marine bacterium, *Zooglea* sp., has been reported to adsorb metal ions like chromium, lead, and iron in solutions (Kong et al., 1998). Biosorption of heavy metals by *E. cloaceae* is reported by Iyer et al. (2004) and Iyer, Mody, and Jha (2005b).

8.5. Enhanced oil recovery

The *in situ* production of xanthan-like polysaccharide in the oil-bearing strata has been suggested as a means of aiding tertiary oil recovery (Wells, 1977). *Volcaniella eurihalina* F2–7 is known to synthesize an exopolysaccharide, the rheological properties of which are stable to pH and inorganic salts, which makes it a suitable candidate for enhanced oil recovery (Calvo, Ferrer, Martínez-Checa, Béjar, & Quesada, 1995). An exopolysaccharide produced by *E. cloaceae* has been reported to have good viscosity even at high temperature, which makes it a probable candidate for microbial enhanced oil recovery (Iyer, Mody, & Jha, 2005a).

9. CONCLUSIONS

Microbes isolated from marine environments offer a great diversity in chemical and physical properties of their EPS compared to anywhere else in the biosphere. Bacteria from remote areas still remain virtually unexplored, and there is no doubt that extreme environments such as deep-sea hydrothermal vents are a rich source of microorganisms of biotechnological importance. A number of interesting and unique polysaccharides have been isolated from these microorganisms and are expected to find applications in the very near future in different industries. Further screenings are underway as well as research into understanding the structure–function relationships of these unusual polymers. Those strains that produce EPS exhibiting novel properties will be transferred to private industries interested in further evaluating the polymers for commercial development. However, despite the interesting properties and subsequent applications of these polysaccharides, other factors including the yield, the price, and the markets for such new molecules will ultimately determine the commercial development of these polymers.

ACKNOWLEDGMENTS

This research was supported by a grant from Marine Bioprocess Research Center of the Marine Biotechnology Program funded by the Ministry of Oceans and Fisheries, Republic of Korea.

REFERENCES

Alldredge, A. L. (2000). Interstitial dissolved organic carbon (DOC) concentrations within sinking marine aggregates and their potential contribution to carbon flux. *Limnology and Oceanography*, *45*, 1245–1253.

Ashtaputre, A., & Shah, A. (1995). Emulsifying property of a viscous exopolysaccharide from *Sphingomonas paucimobilis*. *World Journal of Microbiology and Biotechnology*, *11*(2), 219–222.

Baba, M., Pauwels, R., Balzarini, J., Arnout, J., Desmyter, J., & De Clercq, E. (1998). Mechanism of inhibitory effect of dextran sulfate and heparin on replication of human immunodeficiency virus *in vitro*. *Proceedings of the National Academy of Sciences of the United States of America*, *85*(16), 6132–6136.

Bai, Y., Zhang, P., Chen, G., Cao, J., Huang, T., & Chen, K. (2012). Macrophage immunomodulatory activity of extracellular polysaccharide (PEP) of Antarctic bacterium *Pseudoaltermonas* sp. S-5. *International Immunopharmacology*, *12*(4), 611–617.

Barreto-Bergter, E., & Gorin, P. A. (1983). Structural chemistry of polysaccharides from fungi and lichens. *Advances in Carbohydrate Chemistry and Biochemistry*, *41*, 67–103.

Boyle, C. D., & Reade, A. E. (1983). Characterization of two extracellular polysaccharides from marine bacteria. *Applied and Environmental Microbiology*, *46*(2), 392–399.

Calvo, C., Ferrer, M., Martínez-Checa, F., Béjar, V., & Quesada, E. (1995). Some rheological properties of the extracellular polysaccharide produced by *Volcaniella eurihalina* F2-7. *Applied Biochemistry and Biotechnology*, *55*(1), 45–54.

Chi, Z., Su, C., & Lu, W. (2007). A new exopolysaccharide produced by marine *Cyanothece* sp. 113. *Bioresource Technology*, *98*(6), 1329–1332.

Christensen, B. E., Kjosbakken, J., & Smidsrød, O. (1985). Partial chemical and physical characterization of two extracellular polysaccharides produced by marine, periphytic *Pseudomonas* sp. strain NCMB 2021. *Applied and Environmental Microbiology*, *50*(4), 837–845.

Decho, A. W., & Herndl, G. J. (1995). Microbial activities and the transformation of organic matter within mucilaginous material. *Science of the Total Environment*, *165*(1), 33–42.

De Philippis, R., Margheri, M. C., Materassi, R., & Vincenzini, M. (1998). Potential of unicellular cyanobacteria from saline environments as exopolysaccharide producers. *Applied and Environmental Microbiology, 64*(3), 1130–1132.

De Philippis, R., Sili, C., Paperi, R., & Vincenzini, M. (2001). Exopolysaccharide-producing cyanobacteria and their possible exploitation: A review. *Journal of Applied Phycology, 13*(4), 293–299.

Fang, Y., Ahmed, S., Liu, S., Wang, S., Lu, M., & Jiao, Y. (2013). Optimization of antioxidant exopolysaccharides production by *Bacillus licheniformis* in solid state fermentation. *Carbohydrate Polymers, 98*(2), 1377–1382.

Fang, Y., Liu, S., Lu, M., Jiao, Y., & Wang, S. (2012). A novel method for promoting antioxidant exopolysaccharides production of *Bacillus licheniformis*. *Carbohydrate Polymers, 92*(2), 1172–1176.

Guezennec, J. (2002). Deep-sea hydrothermal vents: A new source of innovative bacterial exopolysaccharides of biotechnological interest. *Journal of Industrial Microbiology and Biotechnology, 29*(4), 204–208.

Guo, S., Mao, W., Han, Y., Zhang, X., Yang, C., Chen, Y., et al. (2010). Structural characteristics and antioxidant activities of the extracellular polysaccharides produced by marine bacterium *Edwardsiella tarda*. *Bioresource Technology, 101*(12), 4729–4732.

Holmström, C., & Kjelleberg, S. (1999). Marine *Pseudoalteromonas* species are associated with higher organisms and produce biologically active extracellular agents. *FEMS Microbiology Ecology, 30*(4), 285–293.

Inoue, T., Murakawa, S., & Endo, A. (1992). Production and partial properties of an extracellular polysaccharide from *Streptomyces* sp. A-1845. *Journal of Fermentation and Bioengineering, 73*(6), 440–442.

Iwabuchi, N., Sunairi, M., Urai, M., Itoh, C., Anzai, H., Nakajima, M., et al. (2002). Extracellular polysaccharides of *Rhodococcus rhodochrous* S-2 stimulate the degradation of aromatic components in crude oil by indigenous marine bacteria. *Applied and Environmental Microbiology, 68*(5), 2337–2343.

Iyer, A., Mody, K., & Jha, B. (2004). Accumulation of hexavalent chromium by an exopolysaccharide producing marine *Enterobacter cloaceae*. *Marine Pollution Bulletin, 49*(11), 974–977.

Iyer, A., Mody, K., & Jha, B. (2005a). Rheological properties of an exopolysaccharide produced by a marine *Enterobacter cloacae*. *National Academy Science Letters, 28*(3–4), 119–123.

Iyer, A., Mody, K., & Jha, B. (2005b). Biosorption of heavy metals by a marine bacterium. *Marine Pollution Bulletin, 50*(3), 340–343.

Iyer, A., Mody, K., & Jha, B. (2006). Emulsifying properties of a marine bacterial exopolysaccharide. *Enzyme and Microbial Technology, 38*(1), 220–222.

Jansson, P., Kenne, L., & Lindberg, B. (1975). Structure of the extracellular polysaccharide from *Xanthomonas campestris*. *Carbohydrate Research, 45*(1), 275–282.

Kang, K. S., Veeder, G. T., Mirrasoul, P. J., Kaneko, T., & Cottrell, I. W. (1982). Agar-like polysaccharide produced by a *Pseudomonas* species: Production and basic properties. *Applied and Environmental Microbiology, 43*(5), 1086–1091.

Kano, K., & Pettitt, D. (1993). Xanthan, gellan, welan, and rhamsan. In R. L. Whistler, & J. N. BeMiller (Eds.), *Polysaccharides and their derivatives* (pp. 341–397). San Diego: Industrial Gums Academic Press.

Kong, J.-Y., Lee, H.-W., Hong, J.-W., Kang, Y.-S., Kim, J.-D., Chang, M.-W., et al. (1998). Utilization of a cell-bound polysaccharide produced by the marine bacterium *Zoogloea* sp.: New biomaterial for metal adsorption and enzyme immobilization. *Journal of Marine Biotechnology, 6*(2), 99–103.

Krembs, C., Eicken, H., Junge, K., & Deming, J. (2002). High concentrations of exopolymeric substances in Arctic winter sea ice: Implications for the polar ocean carbon cycle and cryoprotection of diatoms. *Deep Sea Research Part I: Oceanographic Research Papers, 49*(12), 2163–2181.

Kumar, C. G., Joo, H.-S., Choi, J.-W., Koo, Y.-M., & Chang, C.-S. (2004). Purification and characterization of an extracellular polysaccharide from haloalkalophilic *Bacillus* sp. I-450. *Enzyme and Microbial Technology, 34*(7), 673–681.

Laurienzo, P. (2010). Marine polysaccharides in pharmaceutical applications: An overview. *Marine Drugs, 8*(9), 2435–2465.

Lee, H. K., Chun, J., Moon, E. Y., Ko, S.-H., Lee, D.-S., Lee, H. S., et al. (2001). *Hahella chejuensis* gen. nov., sp. nov., an extracellular-polysaccharide-producing marine bacterium. *International Journal of Systematic and Evolutionary Microbiology, 51*(Pt. 2), 661–666.

Lin, C. C., & Casida, L. (1984). Gelrite as a gelling agent in media for the growth of thermophilic microorganisms. *Applied and Environmental Microbiology, 47*(2), 427–429.

Loaëc, M., Olier, R., & Guezennec, J. (1997). Uptake of lead, cadmium and zinc by a novel bacterial exopolysaccharide. *Water Research, 31*(5), 1171–1179.

Loaëc, M., Olier, R., & Guezennec, J. (1998). Chelating properties of bacterial exopolysaccharides from deep-sea hydrothermal vents. *Carbohydrate Polymers, 35*(1), 65–70.

Logan, B. E., & Hunt, J. R. (1987). Advantages to microbes of growth in permeable aggregates in marine systems. *Limnology and Oceanography, 32*(5), 1034–1048.

Mancuso Nichols, C., Garon, S., Bowman, J., Raguénès, G., & Guezennec, J. (2004). Production of exopolysaccharides by Antarctic marine bacterial isolates. *Journal of Applied Microbiology, 96*(5), 1057–1066.

Manivasagan, P., Sivasankar, P., Venkatesan, J., Senthilkumar, K., Sivakumar, K., & Kim, S.-K. (2013). Production and characterization of an of extracellular polysaccharide from *Streptomyces violaceus* MM72. *International Journal of Biological Macromolecules, 59*, 29–38.

Matsuda, M., Shigeta, S., & Okutani, K. (1999). Antiviral activities of marine *Pseudomonas* polysaccharides and their oversulfated derivatives. *Marine Biotechnology, 1*(1), 68–73.

Morin, A. (1998). Screening of polysaccharide-producing microorganisms, factors influencing the production and recovery of microbial polysaccharides. In S. Dumitriu (Ed.), *Polysaccharides: Structural diversity and functional versatility* (pp. 275–296). New York: Marcel Dekker Inc. Publication.

Nichols, C. M., Guezennec, J., & Bowman, J. (2005). Bacterial exopolysaccharides from extreme marine environments with special consideration of the southern ocean, sea ice, and deep-sea hydrothermal vents: A review. *Marine Biotechnology, 7*(4), 253–271.

Nichols, C. M., Lardière, S. G., Bowman, J. P., Nichols, P. D., Gibson, J. A., & Guézennec, J. (2005). Chemical characterization of exopolysaccharides from Antarctic marine bacteria. *Microbial Ecology, 49*(4), 578–589.

Nishino, T., & Nagumo, T. (1992). Anticoagulant and antithrombin activities of oversulfated fucans. *Carbohydrate Research, 229*(2), 355–362.

Okutani, K. (1984). Antitumor and immunostimulant activities [to mice] of polysaccharide produced by a marine bacterium of the genus Vibrio. *Bulletin of the Japanese Society of Scientific Fisheries, 50*, 1035–1037.

Okutani, K. (1985). Isolation and fractionation of an extracellular polysaccharide from marine *Vibrio*. *Bulletin of the Japanese Society of Scientific Fisheries, 51*, 493–496.

Okutani, K. (1992). Antiviral activities of sulfated derivatives of a fucosamine-containing polysaccharide of marine bacterial origin. *Nippon Suisan Gakkaishi, 58*(5), 927–930.

Raguénès, G., Peres, A., Ruimy, R., Pignet, P., Christen, R., Loaëc, M., et al. (1997). *Alteromonas infernus* sp. nov., a new polysaccharide-producing bacterium isolated from a deep-sea hydrothermal vent. *Journal of Applied Microbiology, 82*(4), 422–430.

Rosenberg, E., Zuckerberg, A., Rubinovitz, C., & Gutnick, D. (1979). Emulsifier of Arthrobacter RAG-1: Isolation and emulsifying properties. *Applied and Environmental Microbiology, 37*(3), 402–408.

Senni, K., Pereira, J., Gueniche, F., Delbarre-Ladrat, C., Sinquin, C., Ratiskol, J., et al. (2011). Marine polysaccharides: A source of bioactive molecules for cell therapy and tissue engineering. *Marine Drugs, 9*(9), 1664–1681.

Shah, V., Garg, N., & Madamwar, D. (1999). Exopolysaccharide production by a marine cyanobacterium *Cyanothece* sp. *Applied Biochemistry and Biotechnology, 82*(2), 81–90.

Shah, V., Ray, A., Garg, N., & Madamwar, D. (2000). Characterization of the extracellular polysaccharide produced by a marine cyanobacterium, *Cyanothece* sp. ATCC 51142, and its exploitation toward metal removal from solutions. *Current Microbiology, 40*(4), 274–278.

Su, C., Chi, Z., & Lu, W. (2007). Optimization of medium and cultivation conditions for enhanced exopolysaccharide yield by marine *Cyanothece* sp. 113. *Chinese Journal of Oceanology and Limnology, 25*, 411–417.

Suresh Kumar, A., Mody, K., & Jha, B. (2007). Bacterial exopolysaccharides—A perception. *Journal of Basic Microbiology, 47*(2), 103–117.

Sutherland, I. W. (1996). Extracellular polysaccharides. In H. J. Rehm & G. Reed (Eds.), *Biotechnology: Vol. 6*. (pp. 615–657). Weinheim: VCH.

Sutherland, I. W. (2001). Microbial polysaccharides from Gram-negative bacteria. *International Dairy Journal, 11*(9), 663–674.

Tzianabos, A. O. (2000). Polysaccharide immunomodulators as therapeutic agents: Structural aspects and biologic function. *Clinical Microbiology Reviews, 13*(4), 523–533.

Vanhooren, P., & Vandamme, E. J. (1998). Biosynthesis, physiological role, use and fermentation process characteristics of bacterial exopolysaccharides. *Recent Research Developments in Fermentation and Bioengineering, 1*, 253–300.

Vanhooren, P., & Vandamme, E. (2000). Microbial production of clavan, an L-fucose rich exopolysaccharide. *Progress in Biotechnology, 17*, 109–114.

Volpi, N., Sandri, I., & Venturelli, T. (1995). Activity of chondroitin ABC lyase and hyaluronidase on free-radical degraded chondroitin sulfate. *Carbohydrate Research, 279*, 193–200.

Weiner, R., Langille, S., & Quintero, E. (1995). Structure, function and immunochemistry of bacterial exopolysaccharides. *Journal of Industrial Microbiology, 15*(4), 339–346.

Wells, J. (1977). Extracellular microbial polysaccharides—A critical overview. In P. A. Sanford, & A. Laskin (Eds.), *ACS symposium series: Vol. 45*. (pp. 299–313). Washington, DC: American Chemical Society.

Wolfaardt, G. M., Lawrence, J. R., & Korber, D. R. (1999). Function of EPS. In J. Wingender, T. R. Neu, & H. C. Flemming (Eds.), *Microbial extracellular polymeric substances* (pp. 171–200). New York, NY: Springer-Verlag.

Wuertz, S., Müller, E., Spaeth, R., Pfleiderer, P., & Flemming, H. (2000). Detection of heavy metals in bacterial biofilms and microbial flocs with the fluorescent complexing agent Newport Green. *Journal of Industrial Microbiology and Biotechnology, 24*(2), 116–123.

CHAPTER SIX

Biological Activities of Alginate

Mikinori Ueno, Tatsuya Oda[1]

Division of Biochemistry, Faculty of Fisheries, Nagasaki University, Nagasaki, Japan
[1]Corresponding author: e-mail address: t-oda@nagasaki-u.ac.jp

Contents

1. Introduction   96
2. Macrophage-Stimulating Activities of Alginates   98
    2.1 Sources of alginates   98
    2.2 Preparation of alginate oligomers   98
    2.3 Cytokine-inducing activities of alginates with different molecular characteristics in mouse macrophage cell line RAW264.7 cells   100
    2.4 Effects of enzymatic digestion on the TNF-α-Inducing activities of alginates   101
    2.5 Effects of specific MAP kinase inhibitors on the TNF-α-inducing activities of alginate (I-S) and its enzymatically digested alginate oligomer mixture   103
    2.6 Nitric oxide-inducing activities of alginate and its enzymatically digested alginate oligomer mixture   105
3. Antioxidant Activities of Alginate   106
References   108

Abstract

To gain insight into the structure–activity relationship of alginate, we examined the effect of alginates with varying molecular weights and *M/G* ratio on murine macrophage cell line, RAW264.7 cells in terms of induction of tumor necrosis factor-α (TNF-α) secretion. Among the alginates tested, alginate with the highest molecular weight (MW 38,000, *M/G* 2.24) showed the most potent TNF-α-inducing activity. Alginates having higher *M/G* ratio tended to show higher activity. These results suggest that molecular size and *M/G* ratio are important structural parameters influencing the TNF-α-inducing activity. Interestingly, enzymatic depolymerization of alginate with bacterial alginate lyase resulted in dramatic increase in the TNF-α-inducing activity. The higher activity of enzymatically digested alginate oligomers to induce nitric oxide production from RAW264.7 cells than alginate polymer was also observed. On the other hand, alginate polymer and oligomer showed nearly equal hydroxyl radical scavenging activities.

1. INTRODUCTION

Alginate is an acidic linear polysaccharide found in certain species of brown algae such as *Macrocystis pyrifera* and *Ascophyllum nodosum*. Alginate consists of α-L-guluronate (G) and β-D-mannuronate (M), and the residues are arranged in a block structure of a homopolymer (polyguluronate or polymannuronate) or a heteropolymer (a mixed sequence of these residues). These block structures are expressed as G-blocks, M-blocks, and MG-blocks (Haug, 1965). As a quite safety ingredient, alginate is utilized in a wide range of commercial food and medical industries as food additive, cosmetic, and pharmaceutical materials inducing thickener, humectant, disintegrator of tablet, and dispersion stabilizers. As an attractive bioactive agent, it has extensively been studied and found that alginate has a number of physiological activities including the antitumor activities in experimental animals (Fujihara & Nagumo, 1992; Iizima-Mizui et al., 1985), enhanced cholesterol excretion and glucose tolerance (Kimura, Watanabe, & Okuda, 1996), inhibition of smooth muscle cell proliferation (Logeart, PrigentRichard, BoissonVidal, et al., 1997; Logeart, PrigentRichard, Jozefonvicz, & Letourneur, 1997), inhibition of histamine release from mast cells (Asada et al., 1997), and inhibition of histidine decarboxylase and nuclear factor-κB activation in *in vivo* and *in vitro* models (Jeong et al., 2006). The differences in the molecular weight, the M/G ratio, and the entire molecular conformation seem to influence the physicochemical properties and bioactivities of alginates (Asada et al., 1997; Jeong et al., 2006; Kimura et al., 1996; Logeart, PrigentRichard, BoissonVidal, et al., 1997; Logeart, PrigentRichard, Jozefonvicz, et al., 1997; Otterlei et al., 1991). For instance, it has been shown that short alginate polymers consisted of M-blocks and MG-blocks stimulated monocytes to produce tumor necrosis factor (TNF)-α, interleukin (IL)-1, and IL-6, whereas alginate consisted of G-blocks did not induce cytokine production (Otterlei et al., 1991).

Since alginates have gentle gelling properties in the presence of divalent cations such as calcium, alginates are also used for live cell encapsulation *in vitro* (Fremond et al., 1993) and *in vivo* (Chang et al., 2001) and for several tissue engineering applications (Atala, Kim, Paige, Vacanti, & Retik, 1994; Hauselmann et al., 1996).

Alginate oligosaccharides (oligomers) with relatively low molecular weight prepared by enzymatic degradation of alginate polymers are also known to have several biological activities such as suppression of fibroblast proliferation and collagen synthesis in human skin (Tajima et al., 1999), stimulation of endothelial cell growth and migration (Kawada, Hiura, Tajima, & Takahara, 1999), stimulation of human keratinocyte growth (Kawada et al., 1997), and suppression of Th2 development and IgE secretion through inducing IL-12 secretion (Yoshida, Hirano, Wada, Takahashi, & Hattori, 2004). Since the alginate oligomers have fairly low viscosity in aqueous solution even at quite high concentration, and have no gel-forming property in the presence of calcium, it is considered that alginate oligomers are more applicable for *in vivo* systems (Iwamoto et al., 2005; Kurachi et al., 2005). In fact, it was found that intraperitoneal (i.p.) administration of enzymatically depolymerized alginate oligomers to mice resulted in significant increase in serum levels of various cytokines such as TNF-α and granulocyte colony-stimulating factor (Yamamoto, Kurachi, Yamaguchi, & Oda, 2007). These findings suggest that alginate oligomers even at mixture are capable to exhibit certain biological activities in both *in vitro* and *in vivo* systems. In addition to these mammalian models, it has been reported that enzymatically depolymerized alginates promote the growth of bifidobacteria, while the original alginate polymer had no such effect (Akiyama et al., 1992).

Alginates seem to have some biological effects on plants as well. Several studies on the effects of alginates on physiological activities of plant cells have been reported (Aoyagi et al., 1996; Aoyagi, Sakamoto, Asada, & Tanaka, 1998; Aoyagi, Yasuhira, & Tanaka, 1999; Tanaka, Kaneko, Aoyagi, Yamamoto, & Fukunaga, 1996; Tanaka, Yamashita, Aoyagi, Yamamoto, & Fukunaga, 1996; Tomida et al., 2010). Alginate oligomers prepared with bacterial alginate lyase increased shoot elongation of komatsuna (*Brassica rapa* var. *pervidis*) seeds (Yonemoto et al., 1993) and promoted the elongation of barley roots (Tomoda, Umemura, & Adachi, 1994). It has recently been reported that pentaguluronate prepared from alginate significantly promoted the growth of rice and carrot root (Xu, Iwamoto, Kitamura, Oda, & Muramatsu, 2003). Furthermore, it has been reported that a mixture of alginate oligomers prepared by digestion of alginate polymer with alginate lyase promotes growth of the freshwater green microalga *Chlamydomonas reinhardtii* in a concentration-dependent manner, whereas a mixture of alginate oligomers obtained by acid hydrolysis had no growth-promoting activity (Yamasaki et al., 2012). Interestingly, the levels of some fatty acids such as C16:0, C18:2 *cis*, and C18:3 *n*-3 increased in *C. reinhardtii* treated with enzyme-digested alginate oligomers. These results suggest the possibility that enzyme-digested alginate oligomers can be used as a promoting agent for efficient biomass production by reducing culture times and by changing cellular fatty acid levels of a green microalga, which might be potentially useful as a source of biofuel.

In this chapter, we describe the biological activities of alginates and their enzymatically depolymerized oligomers mainly focused on their effects on macrophages. Antioxidant property of alginate is also addressed. The structure–activity relationships of alginate polymers and oligomers were discussed based on these activities.

2. MACROPHAGE-STIMULATING ACTIVITIES OF ALGINATES

2.1. Sources of alginates

Sodium alginates are commercially available from several companies. We mainly obtained sodium alginates (food and medical usage grade) with varying viscosities from KIMIKA Co. (Tokyo, Japan) and used without further purification. Other sodium alginates were from Nacalai Tesque Inc. (Kyoto, Japan) and Sigma Chemical (St. Louis, MO, USA). The ratio of mannuronate residues to guluronate residues (M/G ratio) of each alginate was estimated by circular dichroism analysis (Morris, Rees, & Thom, 1980). The molecular weight of each alginate was estimated by the combination of the measurement of total uronic acid content and reducing *endo*-sugar content (Kurachi et al., 2005). The sodium alginates used in this study are listed in Table 6.1.

2.2. Preparation of alginate oligomers

There are mainly two methods for preparing alginate oligomers. A method used enzyme was conducted by basically the same procedure as described previously (Yamamoto et al., 2007). Alginate lyase required for the preparation of alginate oligomers was purified from culture medium of *Pseudoalteromonas* sp. strain No. 272, as described previously (Iwamoto, Hidaka, Oda, & Muramatsu, 2003). This enzyme, which can recognize both polyguluronate and polymannuronate, efficiently produces unsaturates alginate oligomers with various degree of polymerization (DP) (Iwamoto, Hidaka, et al., 2003; Iwamoto, Xu, et al., 2003). Recently, alginate lyase with similar substrate specificity is commercially available from Nagase ChemteX Co. (Osaka, Japan). For the enzymatic digestion, 5% of alginate samples in aqueous solution were incubated with alginate lyase (final 1 µg/ml) at 40 °C for 3 days. The enzymatic reaction was stopped by heating

Table 6.1 TNF-α-inducing activities of various alginates with different molecular sizes and M/G ratios from RAW264.7 cells

Type of alginate[a]	Molecular size	M/G ratio	TNF-α (pg/ml)
I-S	38,000	2.24	$2115 \pm 148^{b,c}$
I-3	31,600	2.34	$1765 \pm 255^{b,c}$
I-5	25,000	2.08	$864 \pm 5^{b,c}$
I-1	24,000	1.50	89 ± 15^{b}
ULV-3	9000	1.86	41 ± 3^{b}
IL-2	21,000	1.71	$644 \pm 68^{b,c}$
Sigma (A 2033)	21,900	2.12	$1156 \pm 37^{b,c}$
Sigma (A 2158)	9500	1.96	$107 \pm 11^{b,c}$
Nacalai (32–75)	21,600	3.17	$1097 \pm 106^{b,c}$
Control	–	–	$5 + 4$

[a] I-S, I-3, I-5, I-1, ULV-3, and IL-2 were alginate polymers obtained from KIMIKA Co. Sigma and Nacalai mean alginate polymers purchased from Sigma Chemical and Nacalai Tesque Inc., respectively.
[b] Significantly different ($p < 0.05$) from the control value.
[c] Significantly different ($p < 0.05$) from the ULV-3 value.
RAW264.7 cells in 96-well plates (2×10^4 cells/well) were incubated with 1000 μg/ml of each sample in RPMI 1640 medium containing 10% FBS at 37 °C. After 24 h, the levels of TNF-α in the cell-free supernatants were measured by ELISA.

the solution in boiling water for 10 min. Gel–filtration analysis suggested that the enzymatically digested product contained several oligomers with DP of 3–9 (i.e., with molecular weight of 529–1585) (Kurachi et al., 2005). Gel-filtration chromatography also suggested that similar composition of oligomer mixture could be prepared by this procedure even from different alginate polymers with different molecular sizes and M/G ratios (Kurachi et al., 2005). Before use, all the samples were filtered through an endotoxin-removing filter (Zetapor Dispo filter) purchased from Wako Pure Chemical Industries, Ltd. (Osaka, Japan). This filtration procedure could reduce even the bioactivity of 1 μg/ml of lipopolysaccharide (LPS) to negligible level.

Acid hydrolysis is also used as another procedure to prepare alginate oligomers. Alginate polymer (1%) in 200 ml of acidic solution (0.3 M HCl, pH 4.0) was treated at 121 °C for 80 min. After cooling down to room temperature, the sample solution was neutralized with 1 M NaOH and lyophilized in a vacuum freeze dryer. By this procedure, saturated alginate oligomer mixtures with free carboxyl groups are obtained (Matsubara, Iwasaki, & Muramatsu, 1998). Previous studies have demonstrated that enzymatically depolymerized unsaturated alginate oligomers tended to exhibit more potent biological activities than saturated alginate oligomers prepared by acid hydrolysis (Iwamoto et al., 2005). Hence, in this chapter, we show the data on enzymatically depolymerized unsaturated alginate oligomers together with alginate polymers.

2.3. Cytokine-inducing activities of alginates with different molecular characteristics in mouse macrophage cell line RAW264.7 cells

To analyze structure–activity relationship of alginates, TNF-α-inducing activities of various alginates with different structural characteristics were examined in RAW264.7 cells (mouse macrophage cell line), which were from the American Type Culture Collection (Rockville, MD, USA), and cultured in RPMI 1640 medium supplemented with 10% fetal bovine serum (FBS), penicillin (100 units/ml), and streptomycin (100 μg/ml) as described previously (Iwamoto et al., 2005). Monolayers of RAW264.7 cells in 96-well plates (2×10^4 cells/well) were cultured with each alginate sample in the growth medium. After 24 h, the supernatant was withdrawn from each well and assayed for TNF-α. The concentration of TNF-α in culture supernatant was estimated by a sandwich enzyme-linked immunosorbent assay (ELISA) as described previously (Iwamoto et al., 2005).

As shown in Table 6.1, I-S, which has the largest molecular size among the alginates tested, showed the highest activity to induce the secretion of TNF-α from RAW264.7 cells. I-3, which has molecular weight above 30,000, also showed relatively high activity. The activity of I-S was concentration dependent, and a significant effect was observed even at 100 μg/ml (Fig. 6.1).

On the other hand, ULV-3 and Sigma alginate (A 2158) with molecular weight <10,000 had no significant activity. These results suggest that there is a tendency that the higher the molecular size, the higher the TNF-α-inducing activity. Other alginate samples (I-5, I-1, IL-2, Sigma-A 2158, and Nacalai) have similar molecular weight (21,000–25,000), but their activities were different significantly. In addition to the molecular size, it seems that the *M/G* ratio of alginates also influences the activity. For instance, the activity of I-5 (*M/G* ratio of 2.08) was 10-fold higher than that of I-1 (*M/G* ratio of 1.50), while I-5 and I-1 have nearly similar molecular size (Table 6.1). Alginates from Sigma (A 2033) and Nacalai, which have relatively high *M/G* ratio (1.96 and 3.17), showed higher activities than that of I-1, even though their molecular sizes were smaller than that of I-1. In

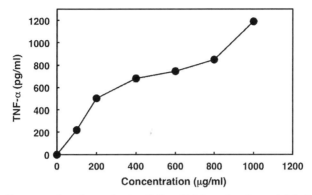

Figure 6.1 Concentration-dependent activity of high-molecular weight alginate polymer (I-S) to induce TNF-α secretion from RAW264.7 cells. RAW264.7 cells in 96-well plates (2×10^4 cells/well) were incubated with varying concentrations (0–1000 μg/ml) of I-S in RPMI 1640 medium containing 10% FBS at 37 °C. After 24 h, the amounts of TNF-α in the culture supernatants were measured by ELISA.

agreement with our findings, Otterlei et al. (1993) have reported that highly mannuronate-enriched polysaccharide isolated from *Pseudomonas aeruginosa* was the most efficient polysaccharides tested in terms of TNF-α induction from human monocytes, and the activity depends strongly on the molecular weight with maximum TNF-α production occurring at molecular weight above 50,000. In addition to the molecular size and M/G ratio, probably the entire molecular conformation of alginate molecules may also be responsible for their biological activities.

2.4. Effects of enzymatic digestion on the TNF-α-inducing activities of alginates

Due to the heterogeneous intramolecular arrangement of M-blocks and G-blocks and existence of M/G random structure in alginate polymers, it is difficult to analyze the exact entire structure of alginates with molecular weight more than 10,000. It is considered that enzymatic depolymerization and subsequent analysis of resulting oligomers may be one way to gain insight into the structure–activity relationship. Therefore, as a pilot study, we examined the effects of enzymatic digestion of alginates on their TNF-α-inducing activities. Interestingly, enzymatic digestion of I-S resulted in dramatic increase in the activity depending on the duration of enzyme treatment (Fig. 6.2).

The activity of I-S increased to nearly 20-fold after 3-day enzyme treatment as compared to the original activity. This enzymatic digestion also

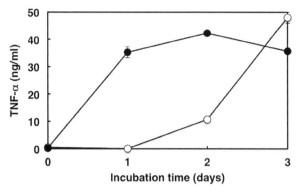

Figure 6.2 Effects of bacterial alginate lyase-mediated enzymatic digestion on the TNF-α-inducing activities of alginates from RAW264.7 cells. To 1% of I-S (○) or 5% of ULV-3 (●) in aqueous solution, alginate lyase at final concentration of 1 μg/ml was added and incubated at 40 °C for the indicated periods of time. After stopping the enzyme activity by heat treatment, samples were passed through endotoxin-removing filter and then subjected to the assay for TNF-α-inducing activity from RAW264.7 cells.

highly enhanced the activity of relatively low-molecular weight alginate sample (ULV-3) that had originally very low TNF-α-inducing activity. The different time course profile between I-S and ULV-3 may be due to their different susceptibility to enzyme digestion. Gel-filtration analysis revealed that the enzymatically digested alginate contained oligomers with various molecular sizes ($DP=3-10$) (Fig. 6.3). Since basically the similar elution profiles of enzymatically digested I-S and ULV-3 were obtained, degradation products derived from I-S and ULV-3 may be similar to each other (Fig. 6.3).

These results suggest that alginate lyase used in this study is capable to convert the alginate polymers to the mixture of alginate oligomers with similar compositions regardless of the molecular size of original alginate polymers. Regarding the effect of enzyme treatment on the biological activities of alginate polymers, it has been reported that alginate oligomers prepared by enzymatic digestion promote growth in bifidobacteria or plants, whereas those alginate polymers without enzymatic treatment have no such effects (Akiyama et al., 1992; Tomoda et al., 1994; Yonemoto et al., 1993). Hence, these findings suggest that enzymatic digestion may lead to increase in the bioactivities of alginate polymers. Furthermore, these findings together with our results suggest that there are some specific oligomers with higher biological activity than original polymers in the enzymatically digested oligomer mixture. In fact, our previous study showed that among the purified unsaturated guluronate (G3–G9) and mannuronate (M3–M9) oligomers, G8 and M7 showed the most potent activities to induce TNF-α secretion from RAW264.7 cells, respectively (Iwamoto et al., 2005).

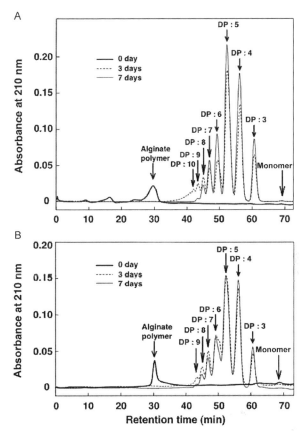

Figure 6.3 Elution profiles of enzymatically digested I-S (A) and ULV-3 (B) on a column of Superdex Peptide 10/300 GL in Waters HPLC system. I-S or ULV-3 was treated with 1 μg/ml of alginate lyase at 40 °C for 0 (—), 3 days (---), and 7 days (—). After heat inactivation of the enzyme activity, each sample was applied to a column of Superdex Peptide 10/300 GL, and elution was monitored at 210 nm. Arrows indicate the estimated degree of polymerization (DP) of each eluted alginate oligomer.

2.5. Effects of specific MAP kinase inhibitors on the TNF-α-inducing activities of alginate (I-S) and its enzymatically digested alginate oligomer mixture

It has been well documented that mitogen-activated protein (MAP) kinases are involved in the secretion of TNF-α responding to extracellular stimulation such as LPS (Carter, Knudtson, Monick, & Hunninghake, 1999). In fact, the alginate (I-S)-induced TNF-α secretion was inhibited with specific MAP kinase inhibitors with different extent depending on the inhibitors.

The inhibition profile of I–S with three inhibitors differed from those of enzymatically digested I–S. Especially, the inhibitory effects of JNK and ERK MAP kinase inhibitors were evidently different between intact I–S and enzymatically digested I–S (Fig. 6.4).

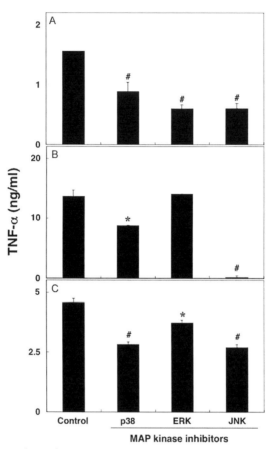

Figure 6.4 Effects of specific MAP kinase inhibitors on the TNF-α-inducing activity of I-S (A), enzymatically digested I-S (B), and lipopolysaccharide (LPS) (C) from RAW264.7 cells. Adherent RAW264.7 cells in 96-well plates (2×10^4 cells/well) were preincubated with 5 μM of p38 (SB 203580), ERK (PD 98059), or JNK (SP 600125)-specific MAP kinase inhibitor for 1 h at 37 C, and then I-S (1 mg/ml), enzymatically digested I-S (1 mg/ml), or LPS (1 ng/ml) was added to the cells. After 24 h, the supernatant was withdrawn from each well and assayed for TNF-α. Each value represents an average of triplicate measurements, and each bar indicates the standard deviation. The values were significantly different from the control value with $^{\#}p < 0.01$ or $^{*}p < 0.05$.

Hence, it seems likely that the underlying mechanisms of cytokine induction in terms of major MAP kinase involved are different between alginate oligomer and polymer. Probably, the recognition sites of RAW264.7 cells linked with signal transduction leading to cytokine production may be different between polymer and oligomer. The inhibition profile of LPS with MAP kinase inhibitors was also different from those of alginate polymer and oligomer.

2.6. Nitric oxide-inducing activities of alginate and its enzymatically digested alginate oligomer mixture

Macrophages play important roles in host defense system, and they produce various inflammatory mediators including cytokines such as TNF-α as described above. In addition to cytokines, stimulated macrophages also secrete nitric oxide (NO) through the activation of inducible nitric oxide synthase (iNOS) from L-arginine (Coleman, 2001; Karpuzoglu & Ahmed, 2006). NO has been known as an essential cytotoxic factor in the killing of pathogens and tumoricidal agent as well (Coleman, 2001; Hajri et al., 1998). It is well documented that a stimulatory pathway in macrophages is initiated by binding of a bacterial cell wall component (LPS) to the CD14-toll-like receptor that triggers a complex kinase cascade and eventually leads to gene activation and subsequent expression of iNOS (Paul, Pendereigh, & Plevin, 1995). It has recently been reported that brown seaweed-derived sulfated polysaccharide fucoidan induces NO production in RAW264.7 cells via p38 MAP kinase and NF-κB-dependent signaling pathways through macrophage scavenger receptors (Nakamura, Suzuki, Wada, Kodama, & Doi, 2006). To estimate NO level in RAW264.7 cells, nitrite, a stable reaction product of NO with molecular oxygen is measured by Griess assay (Jiang, Okimura, Yamaguchi, & Oda, 2011). In brief, adherent RAW264.7 cells in 96-well plates (3×10^4 cells/well) were treated with varying concentrations of polysaccharide samples (0–200 µg/ml) for 24 h in the growth medium at 37 °C, and then the nitrite levels in the culture medium were measured. As shown in Fig. 6.5, alginate polymer induced NO production from RAW264.7 cells in a concentration-dependent manner. Enzymatically depolymerized unsaturated alginate oligomers mixture showed more potent activity than original alginate polymer.

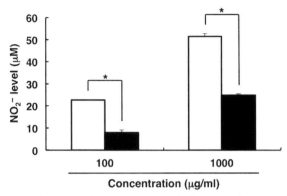

Figure 6.5 Nitric oxide (NO)-inducing activities of alginate polymer and its oligomers mixture in RAW264.7 cells. Adherent RAW264.7 cells (3×10^4 cells/well in 96-well plates) were incubated with the indicated concentrations of alginate polymer (■) and oligomers mixture (□) at 37 °C in the growth medium. After 24 h incubation, the nitrite levels in the culture medium from the treated cells were measured by Griess assay. Each value represents the means ± standard deviation of triplicate measurements. Asterisks indicate significant differences between alginate polymer and alginate oligomer mixture (*$p < 0.01$).

3. ANTIOXIDANT ACTIVITIES OF ALGINATE

Reactive oxygen species (ROS) such as superoxide anion (O_2^-), hydroxyl radical ($^{\bullet}OH$), and hydrogen peroxide (H_2O_2) are produced in various cellular metabolic processes such as respiration, as well as by ultraviolet light, ionizing radiation, and various chemical reactions. In living organisms, ROS levels are controlled by certain enzymes such as superoxide dismutase and antioxidant agents. However, overproduction of ROS can often lead to oxidative stress that in turn causes damage to lipids, proteins, and DNA. Thus, ROS are considered to be involved in a number of pathological conditions including cancer and other various severe diseases (Barnham, Masters, & Bush, 2004; Behl, Davis, Lesley, & Schubert, 1994; Hussain, Hofseth, & Harris, 2003; Reuter, Gupta, Chaturvedi, & Aggarwal, 2010).

The usefulness of polysaccharides and their derivatives in food, agriculture, and medicine has been well documented (Ajithkumar, Andersson, Siika-aho, Tenkanen, & Aman, 2006; Surenjav, Zhang, Xu, Zhang, & Zeng, 2006). Polysaccharides provide various beneficial effects such as lowering blood cholesterol level and blood pressure, and even protective effect on infectious and inflammatory diseases (Hida, Miura, Adachi, & Ohno, 2005). In recent years, considerable attention has directed to marine algae as a rich source of polysaccharides with antioxidant activity (Qi et al., 2005). In addition to the biological activities of alginate polymers and their oligomers described above, recent studies have demonstrated that alginate oligomers (Wang et al., 2007) as well as polymers (Tomida et al., 2010) show potent antioxidant activities. It is well known that hydroxyl radical is the most ROS that can react with various biological molecules, and damaging effect on various biological systems is the strongest among the ROS (Aruoma, Kaur, & Halliwell, 1991; Millic, Dijilas, & Canadanovic-Brunet, 1998). The Fenton reaction is often used as a hydroxyl radical generation system, in which Fe^{2+} and H_2O_2 are reacted to produce hydroxyl radical. Although there are several methods for the detection of hydroxyl radical such as spectrophotometoric or colorimetric methods, electron spin resonance (ESR) method has been used as the most reliable assay for monitoring free radicals because of its high sensitivity and rapidity. Thus, we examined the scavenging activity of alginate on hydroxyl radical by ESR method in this study. When spin-trapping agent DMPO was added to a solution of the Fenton reaction system, the typical 1:2:2:1 ESR signal of the DMPO–OH adduct (an adduct from DMPO and hydroxyl radical) was observed. Figure 6.6 shows the representative ESR spectra of DMPO–OH obtained by the addition of solvent alone or in the presence of alginate polymer or its enzymatically digested alginate oligomers mixture.

In the presence of alginate polymer or alginate oligomers mixture, significant decreases in the height of the second peak of the spectra, which represent relative amount of DMPO–OH adduct, were observed, and the

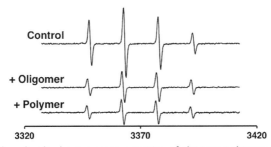

Figure 6.6 Hydroxyl radical scavenging activities of alginate polymer and its enzymatically digested oligomers mixture measured by the ESR-spin-trapping method. The representative ESR spectra of DMPO–OH obtained in the absence (control) or in the presence of 1000 μg/ml of alginate polymer (polymer) or its oligomer mixture (oligomer).

effects of alginate polymer and oligomers mixture were almost similar extent. These results clearly indicate that alginate polymer and its oligomers mixture have an ability to scavenge hydroxyl radicals. Regarding antioxidant activity of polysaccharides, it was proposed that polysaccharides could inhibit the formation of hydroxyl radical, probably due to the hydrogen or electron abstraction mechanism (Chen et al., 2009). It was also reported that the ease of abstraction of the anomeric hydrogen from the internal monosaccharide units made polysaccharides to achieve the scavenging effect (Tsiapali et al., 2001). On the other hand, regarding the hydroxyl radical scavenging mechanism of certain compounds, it has been pointed out that the scavenging activity is not due to direct scavenging but inhibition of hydroxyl radical formation by chelating iron ions in the reaction system (Shon, Kim, & Sung, 2003). In fact, it was reported that molecules that can decrease hydroxyl radical level are those that can chelate iron ions and inhibit the Fenton reaction (Smith, Halliwell, & Aruoma, 1992). However, our previous study clearly indicated that polymannuronate and polygluronate have no chelating activity on Fe^{2+} (Ueno et al., 2012). Similar to our results, comparative study on the radical scavenging activities of alginate oligosaccharides, chitosan oligomers, and fucoidan oligosaccharides with similar molecular weight has demonstrated that alginate oligosaccharides showed the highest hydroxyl radical scavenging activity among these marine oligosaccharides, whereas chitosan oligosaccharides had the highest superoxide radical scavenging activity (Wang et al., 2007). Although the antioxidant mechanism of the oligosaccharides is not fully understood yet, these findings suggest that the chemical structure might be an important factor influencing the antioxidant activity, and alginate oligomer has relatively specific hydroxyl radical scavenging activity.

REFERENCES

Ajithkumar, A., Andersson, R., Siika-aho, M., Tenkanen, M., & Aman, P. (2006). Isolation of cellotriosyl blocks from barley β-glucan with *endo*-1,4-β-glucanase from *Trichoderma reesei*. *Carbohydrate Polymers*, *64*, 233–238.

Akiyama, H., Endo, T., Nakakita, R., Murata, K., Yonemoto, Y., & Okayama, K. (1992). Effect of depolymerized alginates on the growth of bifidobacteria. *Bioscience, Biotechnology, and Biochemistry*, *56*, 355–356.

Aoyagi, H., Okada, M., Akimoto, C., Katsuyama, H., Yoshida, S., Kusakabe, I., et al. (1996). Promotion effect of alginate on chitinase production by *Wasabia japonica*. *Biotechnology Techniques*, *10*, 649–654.

Aoyagi, H., Sakamoto, Y., Asada, M., & Tanaka, H. (1998). Indole alkaloids production by *Catharanthus roseus* protoplasts with artificial cell walls containing of guluronic acid rich alginate gel. *Journal of Fermentation and Bioengineering*, *85*, 306–311.

Aoyagi, H., Yasuhira, J., & Tanaka, H. (1999). Alginate promotes production of various enzymes by *Catharanthus roseus* cells. *Applied Microbiology and Biotechnology*, 52, 429–436.

Aruoma, O. I., Kaur, H., & Halliwell, B. (1991). Oxygen free radicals and human diseases. *Journal of the Royal Society for the Promotion of Health*, 111, 172–177.

Asada, M., Sugie, M., Inoue, M., Nakagomi, K., Hongo, S., Murata, K., et al. (1997). Inhibitory effect of alginic acids on hyaluronidase and on histamine release from mast cells. *Bioscience, Biotechnology, and Biochemistry*, 61, 1030–1032.

Atala, A., Kim, W., Paige, K. T., Vacanti, C. A., & Retik, A. B. (1994). Endoscopic treatment of vesicoureteral reflux with a chondrocyte-alginate suspension. *Journal of Urology*, 152, 641–643.

Barnham, K. J., Masters, C. L., & Bush, A. I. (2004). Neurodegenerative diseases and oxidative stress. *Nature Reviews. Drug Discovery*, 3, 205–214.

Behl, C., Davis, J. B., Lesley, R., & Schubert, D. (1994). Hydrogen peroxide mediates amyloid beta protein toxicity. *Cell*, 77, 817–827.

Carter, A. B., Knudtson, K. L., Monick, M. M., & Hunninghake, G. W. (1999). The p38 mitogen-activated protein kinase is required for NF-κB-dependent gene expression. *The Journal of Biological Chemistry*, 274, 30858–30863.

Chang, S. C. N., Rowley, J. A., Tobias, G., Genes, N. G., Roy, A. K., Mooney, D. J., et al. (2001). Injection molding of chondrocyte/alginate constructs in the shape of facial implants. *Journal of Biomedical Materials Research*, 55, 503–511.

Chen, H., Wang, Z., Qu, Z., Fu, L., Dong, P., & Zhang, X. (2009). Physicochemical characterization and antioxidant activity of a polysaccharide isolated from oolong tea. *European Food Research and Technology*, 229, 629–635.

Coleman, J. W. (2001). Nitric oxide in immunity and inflammation. *International Immunopharmacology*, 1, 1397–1406.

Fremond, B., Malandain, C., Guyomard, C., Chesne, C., Guillouzo, A., & Campion, J. P. (1993). Correction of bilirubin conjugation in the Gunn rat using hepatocytes immobilized in alginate gel beads as an extracorporeal bioartificial liver. *Cell Transplantation*, 2, 453–460.

Fujihara, M., & Nagumo, T. (1992). The effect of the content of D-mannuronic acid and L-guluronic acid blocks in alginates on antitumor activity. *Carbohydrate Research*, 224, 343–347.

Hajri, A., Metzger, E., Vallat, F., Coffy, S., Flatter, E., Evrard, S., et al. (1998). Role of nitric oxide in pancreatic tumour growth: *In vivo* and *in vitro* studies. *British Journal of Cancer*, 78, 841–849.

Haug, A. (1965). Isolation and fractionation with potassium chloride and manganous ions. *Methods in Carbohydrate Chemistry*, 5, 69–73.

Hauselmann, H. J., Masuda, K., Hunziker, E. B., Neidhart, M., Mok, S. S., Michel, B. A., et al. (1996). Adult human chondrocytes cultured in alginate form a matrix similar to native human articular cartilage. *American Journal of Physiology*, 271, C742–C752.

Hida, S., Miura, N. N., Adachi, Y., & Ohno, N. (2005). Effect of *Candida albicans* cell wall glucan as adjuvant for induction of autoimmune arthritis in mice. *Journal of Autoimmunity*, 25, 93–101.

Hussain, S. P., Hofseth, L. J., & Harris, C. C. (2003). Radical causes of cancer. *Nature Reviews. Cancer*, 3, 276–285.

Iizima-Mizui, N., Fujihara, M., Himeno, J., Komiyama, K., Umezawa, I., & Nagumo, T. (1985). Antitumor activity of polysaccharide fractions from the brown seaweed *Sargassum kjelimanianum*. *Kitasato Archives of Experimental Medicine*, 58, 59–71.

Iwamoto, Y., Hidaka, H., Oda, T., & Muramatsu, T. (2003). A study of tryptophan fluorescence quenching of bifunctional alginate lyase from a marine bacterium *Pseudoalteromonas* sp. strain No. 272 by acrylamide. *Bioscience, Biotechnology, and Biochemistry*, 67, 1990–1992.

Iwamoto, M., Kurachi, M., Nakashima, T., Kim, D., Yamaguchi, K., Oda, T., et al. (2005). Structure-activity relationship of alginate oligosaccharides in the induction of cytokine production from RAW264.7 cells. *FEBS Letters, 579,* 4423–4429.

Iwamoto, Y., Xu, X., Tamura, T., Oda, T., & Muramatsu, T. (2003). Enzymatically depolymerized alginate oligomers that cause cytotoxic cytokine production in human mononuclear cell. *Bioscience, Biotechnology, and Biochemistry, 67,* 258–263.

Jeong, H. J., Lee, S. A., Moon, P. D., Na, H. J., Park, R. K., Um, J. Y., et al. (2006). Alginic acid has anti-anaphylactic effects and inhibits inflammatory cytokine expression via suppression of nuclear factor-κB activation. *Clinical and Experimental Allergy, 36,* 785–794.

Jiang, Z., Okimura, T., Yamaguchi, K., & Oda, T. (2011). The potent activity of sulfated polysaccharide, ascophyllan, isolated from *Ascophyllum nodosum* to induce nitric oxide and cytokine production from mouse macrophage RAW264.7 cells: Comparison between ascophyllan and fucoidan. *Nitric Oxide, 25,* 407–415.

Karpuzoglu, E., & Ahmed, S. A. (2006). Estrogen regulation of nitric oxide and inducible nitric oxide synthase (iNOS) in immune cells: Implications for immunity, autoimmune diseases, and apoptosis. *Nitric Oxide, 15,* 177–186.

Kawada, A., Hiura, N., Shiraiwa, M., Tajima, S., Hiruma, M., Hara, K., et al. (1997). Stimulation of human keratinocyte growth by alginate oligosaccharides, a possible co-factor for epidermal growth factor in cell culture. *FEBS Letters, 408,* 43–46.

Kawada, A., Hiura, N., Tajima, S., & Takahara, H. (1999). Alginate oligosaccharides stimulate VEGF-mediated growth and migration of human endothelial cells. *Archives of Dermatological Research, 291,* 542–547.

Kimura, Y., Watanabe, K., & Okuda, H. (1996). Effects of soluble sodium alginate on cholesterol excretion and glucose tolerance in rats. *Journal of Ethnopharmacology, 54,* 47–54.

Kurachi, M., Nakashima, T., Miyajima, C., Iwamoto, Y., Muramatsu, T., Yamaguchi, K., et al. (2005). Comparison of the activities of various alginates to induce TNF-alpha secretion in RAW264.7 cells. *Journal of Infection and Chemotherapy, 11,* 199–203.

Logeart, D., PrigentRichard, S., BoissonVidal, C., Chaubert, F., Durand, P., Jozefonvicz, J., et al. (1997). Fucans, sulfated polysaccharides extracted from brown seaweeds, inhibit vascular smooth muscle cell proliferation. II. Degradation and molecular weight effect. *European Journal of Cell Biology, 74,* 385–390.

Logeart, D., PrigentRichard, S., Jozefonvicz, J., & Letourneur, D. (1997). Fucans, sulfated polysaccharides extracted from brown seaweeds, inhibit vascular smooth muscle cell proliferation. I. Comparison with heparin for antiproliferative activity, binding and internalization. *European Journal of Cell Biology, 74,* 376–384.

Matsubara, Y., Iwasaki, K., & Muramatsu, T. (1998). Action of poly (α-L-guluronate) lyase from *Corynebacterium* sp. ALY-1 strain on saturated oligoguluronates. *Bioscience, Biotechnology, and Biochemistry, 62,* 1055–1060.

Millic, B. L., Dijilas, S. M., & Canadanovic-Brunet, J. M. (1998). Antioxidative activity of phenolic compounds on the metal-ion breakdown of lipid peroxidation system. *Food Chemistry, 61,* 443–447.

Morris, E., Rees, D. A., & Thom, D. (1980). Characterisation of alginate composition and block-structure by circular dichroism. *Carbohydrate Research, 81,* 305–314.

Nakamura, T., Suzuki, H., Wada, Y., Kodama, T., & Doi, T. (2006). Fucoidan induces nitric oxide production via p38 mitogen-activated protein kinase and NF-κB-dependent signaling pathways through macrophage scavenger receptors. *Biochemical and Biophysical Research Communications, 343,* 286–294.

Otterlei, M., Østgaard, K., Skjåk-Baæk, G., Smidsrød, O., Soon-Shiong, P., & Espevik, T. (1991). Induction of cytokine production from human monocytes stimulated with alginate. *Journal of Immunotherapy, 10,* 286–291.

Otterlei, M., Sundan, A., Skjåk-Bræk, G., Ryan, L., Smidsrød, O., & Espevik, T. (1993). Similar mechanisms of action of defined polysaccharides and lipopolysaccharides:

Characterization of binding and tumor necrosis factor alpha induction. *Infection and Immunity, 61*, 1917–1925.

Paul, A., Pendereigh, R. H., & Plevin, R. (1995). Protein kinase C and tyrosine kinase pathways regulate lipopolysaccharide-induced nitric oxide synthase activity in RAW 264.7 murine macrophages. *British Journal of Pharmacology, 114*, 482–488.

Qi, H., Zhang, Q., Zhao, T., Chen, R., Zhang, H., Niu, X., et al. (2005). Antioxidant activities of different sulfate content derivatives of polysaccharide extracted from *Ulva pertusa* (Chlorophyta) in vitro. *International Journal of Biological Macromolecules, 37*, 195–199.

Reuter, S., Gupta, S. C., Chaturvedi, M. M., & Aggarwal, B. B. (2010). Oxidative stress, inflammation, and cancer: How are they linked? *Free Radical Biology and Medicine, 49*, 1603–1616.

Shon, M. Y., Kim, T. H., & Sung, N. J. (2003). Antioxidants and free radical scavenging activity of *Phellinus* (*Phellinus of Hymenochaetaceae*) extracts. *Food Chemistry, 82*, 593–597.

Smith, C., Halliwell, B., & Aruoma, O. I. (1992). Protection by albumin against the prooxidant actions of phenolic dietary components. *Food and Chemical Toxicology, 30*, 483–489.

Surenjav, U., Zhang, L. N., Xu, X. J., Zhang, X. F., & Zeng, F. B. (2006). Effects of molecular structure on antitumor activities of $(1 \rightarrow 3)$-β-D-glucans from different *Lentinus Edodes*. *Carbohydrate Polymers, 63*, 97–104.

Tajima, S., Inoue, H., Kawada, A., Ishibashi, A., Takahara, H., & Hiura, N. (1999). Alginate oligosaccharides modulate cell morphology, cell proliferation and collagen expression in human skin fibroblasts in vitro. *Archives of Dermatological Research, 291*, 432–436.

Tanaka, H., Kaneko, Y., Aoyagi, H., Yamamoto, Y., & Fukunaga, Y. (1996). Efficient production of chitinase by immobilized *Wasabia japonica* cells in double-layered gel fibers. *Journal of Fermentation and Bioengineering, 81*, 220–225.

Tanaka, H., Yamashita, T., Aoyagi, H., Yamamoto, Y., & Fukunaga, Y. (1996). Efficient production of chitinase by immobilized *Wasabia japonica* protoplasts immobilized in double-layered gel fibers. *Journal of Fermentation and Bioengineering, 81*, 394–399.

Tomida, H., Yasufuku, T., Fujii, T., Kondo, Y., Kai, T., & Anraku, M. (2010). Polysaccharides as potential antioxidative compounds for extended-release matrix tablets. *Carbohydrate Research, 345*, 82–86.

Tomoda, Y., Umemura, K., & Adachi, T. (1994). Promotion of barley root elongation under hypoxic conditions by alginate lyase-lysate (A.L.L.). *Bioscience, Biotechnology, and Biochemistry, 58*, 202–203.

Tsiapali, E., Whaley, S., Kalbfleisch, J., Ensley, H. E., Browder, W., & Williams, D. L. (2001). Glucans exhibit weak antioxidant activity, but simulate macrophage free radical activity. *Free Radical Biology and Medicine, 30*, 393–402.

Ueno, M., Hiroki, T., Takeshita, S., Jiang, Z., Kim, D., Yamaguchi, K., et al. (2012). Comparative study on antioxidative and macrophage-stimulating activities of polyguluronic acid (PG) and polymannuronic acid (PM) prepared from alginate. *Carbohydrate Research, 352*, 88–93.

Wang, P., Jiang, X., Jiang, Y., Hu, X., Mou, H., Li, M., et al. (2007). In vitro antioxidative activities of three marine oligosaccharides. *Natural Product Research, 21*, 646–654.

Xu, X., Iwamoto, Y., Kitamura, Y., Oda, T., & Muramatsu, T. (2003). Root growth-promoting activity of unsaturated oligomeric urinates from alginate on carrot and rice plants. *Bioscience, Biotechnology, and Biochemistry, 67*, 2022–2025.

Yamamoto, Y., Kurachi, M., Yamaguchi, K., & Oda, T. (2007). Stimulation of multiple cytokines production in mice by alginate oligosaccharides following intraperitoneal administration. *Carbohydrate Research, 342*, 1133–1137.

Yamasaki, Y., Yokose, T., Nishikawa, T., Kim, D., Jiang, Z., Yamaguchi, K., et al. (2012). Effects of alginate oligosaccharide mixtures on growth and fatty acid composition of the green alga *Chlamydomonas reinhardtii*. *Journal of Bioscience and Bioengineering, 113*, 112–116.

Yonemoto, Y., Tanaka, H., Yamashita, T., Kitabatake, N., Ishida, Y., Kimura, A., et al. (1993). Promotion of germination and shoot elongation of some plants by alginate oligomers prepared with bacterial alginate lyase. *Journal of Fermentation and Bioengineering, 75*, 68–70.

Yoshida, T., Hirano, A., Wada, H., Takahashi, K., & Hattori, M. (2004). Alginic acid oligosaccharide suppresses Th2 development and IgE production by inducing IL-12 production. *International Archives of Allergy and Immunology, 133*, 239–247.

CHAPTER SEVEN

Biological Activities of Carrageenan

Ratih Pangestuti*, Se-Kwon Kim[†,1]

*Research Center for Oceanography, The Indonesian Institute of Sciences, Jakarta, Republic of Indonesia
[†]Department of Marine-bio Convergence Science, Specialized Graduate School Science and Technology Convergence, Marine Bioprocess Research Center, Pukyong National University, Busan, South Korea
[1]Corresponding author: e-mail address: sknkim@pknu.ac.kr

Contents

1. Introduction 113
2. Carrageenan Source and Extraction 115
3. Biological Activities 117
   3.1 Anticoagulant activity 117
   3.2 Antiviral agents 117
   3.3 Cholesterol-lowering effects 119
   3.4 Immunomodulatory activity 119
   3.5 Antioxidant activity 120
   3.6 Antitumor activity 120
4. Food and Technological Applications of Carrageenan 121
5. Conclusions 122
References 123

Abstract

Red seaweeds are popular and economically important worldwide and also well known for their medicinal effects due to the presence of phycocolloids. Carrageenans, the major phycocolloid group of red algae, have been extensively investigated for their vast array of bioactivities such as anticoagulant, antiviral, cholesterol-lowering effects, immunomodulatory activity, and antioxidant. Carrageenan possesses promising activity both *in vitro* and *in vivo*, showing promising potential to be developed as therapeutic agents. In this chapter, attempts have been made to examine the health benefit effects of carrageenans.

1. INTRODUCTION

Algae are a heterogeneous group of plants with a long fossil history. They are very simple, chlorophyll-containing organisms composed of one cell or grouped together in colonies or as organisms with many cells, sometimes collaborating together as simple tissues. Two major types of algae can be identified: the first one are seaweeds or sometimes referred as macroalgae, and the second are microalgae which are found in both benthic and littoral habitats and also throughout the ocean waters. Further, seaweeds can be classified into three classes (red, brown, and green seaweeds) based on their pigmentation. Cell walls of seaweed classes vary with regard to the presence of peculiar amount of structural polysaccharides. For example, the main part of red seaweeds cell walls is represented by sulfated galactans, which are known as carrageenans and agars. These polysaccharides are made up of a repeating disaccharide backbone of 3-linked β-D-galactopyranose (G units) and 4-linked α-D-galactopyranose (D units) or 3,6-anhydrogalactose (DA units) in alternation. They play several functions in seaweeds, such as the energy sources, formation of cell wall, outer capsules, and internal matrices, and provide selective cation absorption.

Agars and carrageenan are important phycocolloids extracted from different red seaweed species. The origin of the name agar comes from the word "agar," which means jelly in the Indonesian and Malay language (*agar–agar*). Meanwhile, the word carrageenan was originated from a small village, Carragheen, on the Irish coast, where the carrageenan-bearing seaweed *Chondrus crispus* or "Irish moss" grows. Nevertheless, economic significance of red seaweeds lies in the utilization as raw material for the production of agar and carrageenan, which are used in foods, cosmetics, medicines, and pharmacy industries. Currently, European seaweed industry relies on phycocolloid production industries with the exception of carrageenan productions. There are about 20 different carrageenan repeating units have been identified on the basis of the sulfation pattern anhydrogalactose content and of the substitution by methoxyl, glycosyl, and/or pyruvate groups. The commercial carrageenans are normally categorized into three main types, which are:

– Kappa (κ) carrageenan forms strong, rigid gels in the presence of potassium ions.
– Iota (ι) carrageenan forms soft, clear, and elastic gels in the presence of calcium ions.
– Lambda (λ) carrageenan does not form gel and normally used to thicken dairy products.

The corresponding IUPAC nomenclature allotted names to Greek symbols such as κ-carrageenan (carrabiose 4′-sulfate, DA-G4S), ι-carrageenan (carrabiose 2,4′-disulfate, DA2SG4S), and λ-carrageenan (carrabiose 2,6,2′-trisulfate, D2S6SG2S).

2. CARRAGEENAN SOURCE AND EXTRACTION

Carrageenan biosynthesis depends on two factors, which can be divided into exogenous, determined by the conditions of growing algae—light, temperature, and salinity; and endogenous, related to the physiology of algae, in particular the species affiliation and stage of development. Endogenous factor is particularly important because red seaweeds have a complex life cycle involving the alternation of vegetative, sexual, and asexual reproduction (Barabanova & Yermak, 2012). *Solieria chordalis* which is present in large quantities in the Brittany shoreline constitutes an interesting material for investigations about the metabolic pathway of carrageenan biosynthesis and the chemical structure of this polysaccharide.

The world manufacturing of carrageenan reach 15,500 tons a year in 1990s and exceeded 50,000 tons in 2007/2008 with a value over US$ 600 million (Guiry, 2008). The main red seaweeds that are used for carrageenan production comprise *C. crispus*, *C. ocellatus*, *Gigartina stellata*, *G. radula*, *G. acicularis*, *G. skottsbergii*, *Eucheuma cottonii*, *E. spinosum*, *Furcellaria fastigata*, and *Hypnea muscifonnis*. In Indonesia and Philippines, the main seaweed species used in the commercial production of carrageenan include *E. cottonii* and *E. spinosum*. These are spiny bushy plants, about 50 cm high, which grow on reefs and shallow lagoons around the Indonesia and Philippines and other tropical coasts. *E. cottonii* yields κ-carrageenan and *E. spinosum* contains ι-carrageenan. Meanwhile, the basic commercial reserves of carrageenophytes in Chile are represented by *G. radula* and *G. skottsbergii*.

Carrageenan is obtained industrially by treating the carrageenophytes with alkali and water to eliminate the soluble compounds. The insoluble residue consists mainly of carrageenan and cellulose. This dry residue is called semi-refined carrageenan. Refined carrageenan is produced by heating the carrageenophytes for several hours with water mixed with an alkali solution. The carrageenan extracted with this process is then precipitated with potassium chloride or with alcohol. The precipitate is separated, dried, and milled, obtaining in this way the refined carrageenan. The manufacturing processes for carrageenan extraction are shown in Fig. 7.1. The processes commence with the selection of seaweed to ensure it is harvested at the right time. After collected, seaweeds were washed to remove sand and stones and then dried quickly to prevent microbial degradation and thus preserve the quality of carrageenan.

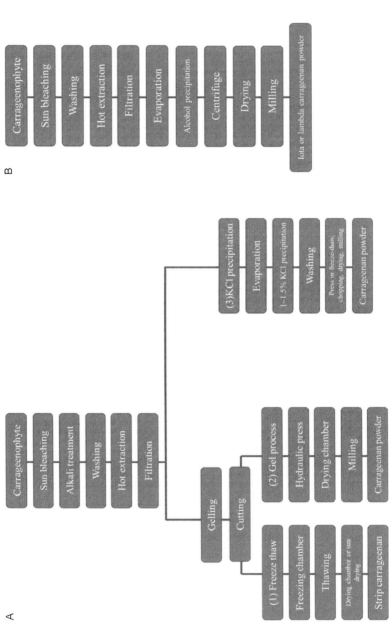

Figure 7.1 Production methods of carrageenans: (A) κ-carrageenans and (B) ι-carrageenans and λ-carrageenans.

3. BIOLOGICAL ACTIVITIES

Carrageenan is having diverse biological and activities including immunomodulatory, anticoagulant, antithrombotic, antiviral, and antitumor effects.

3.1. Anticoagulant activity

The blood coagulation system consists of intrinsic and extrinsic pathways, where a series of factors involve in the mechanism. Blood coagulation is preceded by coagulation factors in order to stop the flow of blood through the injured vessel wall, whenever an abnormal vascular condition and exposure to nonendothelial surfaces at sites of vascular injury occurred. As endogenous or exogenous anticoagulants interfered with the coagulation factors to inactivate or restrict, the blood coagulation can be prolonged or stopped. These anticoagulants are used in therapeutic purposes; hence, the importance of anticoagulants is becoming increasingly recognized. Other than the natural anticoagulants found in the coagulation cascade, a number of anticoagulant materials have also been identified from diverse natural sources. After the investigation of blood anticoagulant properties from red seaweeds, carrageenans are identified with high-anticoagulant activity (Briones et al., 2000; Ermak & Khotimchenko, 1998). Many reports exist on the anticoagulant activity of carrageenan. The most active carrageenan has approximately one-fifteenth the activity of heparin. The principal basis of the anticoagulant activity of carrageenan appears to be an antithrombic property. The λ-carrageenan showed greater antithrombic activity than κ-carrageenan due to its higher sulfate content (Necas & Bartosikova, 2013). The mechanism underlying the anticoagulant activity of carrageenan involves thrombin inhibition.

3.2. Antiviral agents

Sulfated polysaccharides extracted from seaweeds show the antiviral activity against infectious diseases. In addition, red seaweeds-derived sulfated polysaccharides such as carrageenans have inhibitory effects on the entry of enveloped viruses including herpes and human immunodeficiency virus (HIV) into cells. These negatively charged molecules exert their inhibitory effect by interacting with the positive charges on the virus or on the cell surface and thereby prevent the penetration of the virus into the host cells

(Wijesekara, Pangestuti, & Kim, 2010). Carrageenans are able to inhibit the synthesis of viral proteins inside the cells. Furthermore, the presence of sulfate group is necessary for the anti-HIV activity and potency increases with the degree of sulfation, which leads to a hypothesis that anionic charges on the sulfate groups may be effective in inhibiting reverse transcriptase enzyme activity of the virus. However, its strong anticoagulant activity is considered to be an adverse reaction when used as a therapeutic drug for AIDS (Necas & Bartosikova, 2013).

Carrageenan has also been shown to bear antihuman papillomavirus (anti-HPV) activity *in vitro*. Buck et al. noted that carrageenan particularly ι-carrageenan inhibits HPV three orders magnitude more potent than heparin, a highly effective model for HPV inhibitor (Buck et al., 2006). Carrageenan acts primarily by preventing the binding of HPV virions to cells and block HPV infection through a second, postattachment heparin sulfate-independent effect. Those mechanism is consistent by the fact that carrageenan resembles heparin sulfate, which is known as HPV-cell attachment factor. Furthermore, some of milk-based products were able to block HPV infectivity *in vitro*, even when diluted millionfold. In another study, carrageenan has been reported to inhibit genital transmission of HPV in female mouse model of cervicovaginal. Carrageenan was able to generate antigen-specific immune responses and antitumor effects in female (C57BL/6) mice vaccinated with HPV-16 E7 peptide vaccine (Kim & Pangestuti, 2011).

Carrageenans have been recently identified as potent and selective inhibitors of herpes simplex virus (HSV)-1 and HSV-2. Carrageenans isolated from *Meristiella gelidium* were found to be among the most potent sulfated polysaccharides obtained from red seaweeds according to their inhibitory activity against HSV. The HSV inhibition by carrageenan has been investigated in the model of murine infection (Carlucci, Scolaro, Noseda, Cerezo, & Damonte, 2004; Pujol, Carlucci, Matulewicz, & Damonte, 2007; Pujol et al., 2006). The κ-carrageenan extracted from the red seaweed *G. skottsbergii* revealed 100% protection against HSV-2 mortality and replication in a very strict model of murine infection at a high dose of virus. Furthermore, virus or neutralizing antibodies against HSV-2 were not detected in serum of κ-carrageenan-treated animals until 3 weeks after infection. These evidences warrant the availability of κ-carrageenan to protect the whole infectable surface of the mouse vagina.

More recently, Leibbrandt et al. (2010) tested a commercially available nasal spray containing ι-carrageenan in influenza A mouse infection model. Treatment of mice infected with a lethal dose of influenza A PR8/34 H1N1 virus with ι-carrageenan up to 48 h postinfection resulted in a strong protection of mice similar to mice treated with oseltamivir (Leibbrandt et al., 2010). Furthermore, Eccles et al. (2010) investigated the efficacy and safety of a ι-carrageenan nasal spray in patients with common cold symptoms. They concluded that nasal sprays appear to be a promising treatment for safe and effective treatment of early symptoms of the common cold (Eccles et al., 2010).

3.3. Cholesterol-lowering effects

Soluble dietary fiber is defined as dispersible in water, thereby giving rise to viscous gels in the gastrointestinal tract (Yuan, 2007). Seaweed extracts, such as carrageenan, are one of the sources of soluble dietary fiber (Panlasigui, Baello, Dimatangal, & Dumelod, 2003). The biological activity of carrageenan in gastrointestinal tract is associated with viscosity-mediated effects such as reduction in plasma cholesterol and glucose levels. The increased viscosity of the intestinal contents and the decreased rate of digestion and absorption also result in slowing down the diffusion of enzymes, substrates, and nutrients to the absorptive phase, which then results in the lower levels of nutrients, including cholesterol, after a meal, manifested as hypocholesterolemic effect. Reduction of serum cholesterol and lipids may also be due to its ability to bind cholesterol-rich bile in the lower small intestines, resulting in lesser cholesterol absorbed. By the bile–carrageenan binding, less bile is reabsorbed. Serum cholesterol further decreases as cholesterol is used by the liver to replenish the bile acid pool. Fermentation of carrageenan in the large intestines also decreases the biosynthesis of cholesterol. Propionate, one of the short-chain fatty acids produced by the fermentation of carrageenan, inhibits the rate by which the liver produces the body's cholesterol. Hence, carrageenan may also be used to control the cholesterol content in foods because of its ability to mimic the texture and sensory qualities of fat, thereby reducing the total amount of fat in food.

3.4. Immunomodulatory activity

Carrageenans also possess both immunopotentiative and immunosuppressive actions (Yermak & Khotimchenko, 2003). Carrageenan, known to be a potent inflammatory agent in rodents, primes mice leukocytes to produce tumor necrosis factor-α in response to bacterial lipopolysaccharide (Blanque, Meakin, Millet, & Gardner, 1998; Torsdottir, Alpsten, Holm, Sandberg, & Tölli, 1991). However, in some *in vivo* assays, carrageenans also act as inhibitor of macrophage function (Takeyama et al., 1987). Carrageenan may have potential biomedical applications in stimulating the immune system or in controlling macrophage activity to reduce associated negative effects.

3.5. Antioxidant activity

Antioxidant activity of carrageenans has been determined by various methods such as reducing power, iron ion chelation, and total antioxidant activity using beta-carotene–linoleic acid system, 1,1-diphenyl-2-picrylhydrazyl radical scavenging, lipid peroxide inhibition, ferric reducing antioxidant power, nitric oxide scavenging, ABTS radical scavenging, superoxide radical, and hydroxyl radical scavenging assays (Sokolova et al., 2011; Yuan et al., 2006, 2005). In terms of inhibition of superoxide radical and hydroxyl radical formation, λ-carrageenan has relatively stronger inhibitory activity compared with κ-carrageenan and ι-carrageenan (de Souza et al., 2007). Furthermore, it has also been demonstrated that carrageenans with lower molecular weights had better antioxidant activity (Sun, Tao, Xie, Zhang, & Xu, 2010). Sokolova et al. (2011) reported that carrageenans were able to stimulate catalytic activity of SOD from donor erythrocyte.

3.6. Antitumor activity

Several studies have reported that carrageenans have antiproliferative activity in cancer cell lines *in vitro*, as well as inhibitory activity of tumor growth *in vivo* (Haijin, Xiaolu, & Huashi, 2003). In addition, they have antimetastatic activity by blocking the interactions between cancer cells and the basement membrane, inhibiting tumor cell proliferation and tumor cell adhesion to various substrates. However, their exact mechanisms of action are not yet completely understood. Haijin et al. (2003) demonstrated that oral administration of carrageenan oligosaccharide with a molecular weight of 1726 at a dose of 100 mg kg^{-1} markedly inhibited tumor formation in mouse. Antitumor activity of carrageenan might be closely related to their molecular weight. When λ-carrageenans were hydrolyzed with microwave, it significantly increased antitumor activity. Moreover, low-molecular-weight carrageenan showed little improvement of antitumor activity. This suggests that antitumor activity of carrageenan could be significantly enhanced by lowering their molecular weight (Zhou, Sheng, Yao, & Wang, 2006; Zhou et al., 2004, 2005). In addition, carrageenan antitumor activity might have some relevance with the activation of the immunocompetence of the body (Zhou et al., 2004). Recently, Yuan et al. reported that treatment with different κ-carrageenan oligosaccharide derivatives in mice resulted in an increase in tumor inhibition rate and macrophage phagocytosis and cellular immunity, especially on spleen lymphocyte proliferation. The sulfated derivative at the dose 200 µg g^{-1} per day showed the highest antitumor activities, which were significantly higher than the unmodified oligosaccharides (Yuan, Song, Li, Li, & Liu, 2011). It suggested that chemical modification (especially sulfation) of carrageenan oligosaccharides enhances their antitumor effect and boost their antitumor immunity.

4. FOOD AND TECHNOLOGICAL APPLICATIONS OF CARRAGEENAN

Carrageenans are widely utilized in food industry due to their excellent physical functional properties, such as thickening, gelling, and stabilizing abilities. Carrageenans have been considered as safe, as constituents of food products for many years. The safety of carrageenans for use in foods was confirmed in 2001 at the 57th meeting of the Joint Food and Agriculture Organization of the United Nations World Health Organization Expert Committee in Food Additives (JECFA). According to the JECFA, only degraded carrageenans were associated to adverse effects and should not be used as food additives.

Carrageenans are versatile as food additives, as they are capable of binding water, promoting gel formation, and acting as thickeners and stabilizing agents; further advantages lie in the improvement of palatability and appearance. The importance of carrageenans in the food industry derives mainly from the ability of κ- and ι-carrageenan to form elastic gels in the presence of certain cations. Carrageenan is variably called an emulsifier, stabilizer, colloid, or gum. Many products such as soymilk, chocolate, and other flavored milks, dairy products, infant formulas, and nutritional supplement beverages rely upon carrageenan for their uniform consistencies. Large number of dairy products could not be made, packaged, and stored for long periods of time without these sulfated polysaccharides. Carrageenans are used to gel, thicken, or suspend; therefore, they are used in emulsion stabilization, for syneresis control, and for bodying, binding, and dispersion. The major uses of carrageenan are in foods, particularly dairy applications.

Carrageenan is not as widely used in pharmaceutical products as agar, nor has as many industrial applications as algin. Nevertheless, it is employed in the production of toothpastes, dental powders, and insoluble drug preparations, in industrial suspensions (e.g., water paints), and as a stiffener and binder in textile and leather manufacturers. In toothpastes, carrageenans function as a "binder" to impart the desired rheological properties to the paste and to provide the cosmetic quality of "sheen." Toothpastes consist of ingredients which interact in complex and poorly understood ways and the carrageenan often must be carefully tailored to achieve satisfactory performance in a particular formulation.

Under the form of carrageenans, seaweeds are widely used in cosmetology and in thalassotherapy. Due to the ability to increase the viscosity of the solutions and play the role of gelling agents and stabilizers, carrageenan are widely used in various sticks, creams, soaps, shampoos, lotions, foams, and gels. The efficiency of carrageenans in cosmetic preparations has been often claimed with a large variety of biological activities: tonifying, cleaning, hydrating, revitalizing, and antioxidant.

Furthermore, carrageenan also has long been used in drug delivery system. Carrageenans were used to reduce amount of polymorphic transformation in tableting to produce controlled release matrix tablets, to achieve interactions with drugs for modified release systems (Bornhöft, Thommes, & Kleinebudde, 2005; Ellis, Keppeler, & Jacquier, 2009; Keppeler, Ellis, & Jacquier, 2009; Thommes & Kleinebudde, 2006). The area of immobilized cells and organelles has expanded very fast. Many new techniques for the preparation of immobilized cells have been developed during the past decade and a broad spectrum of applications has been investigated. Increasingly, gentler immobilization procedures have evolved to the point when it now seems possible to hold any cell structure and keep it alive and viable. Carrageenan is especially useful for trapping seeds, cells, and microorganisms with or without nutrients and other active materials (Tsen, Lin, & An-Erl King, 2004).

5. CONCLUSIONS

Carrageenans are commercially important phycocolloids. Economic significance of red seaweeds lies in the utilization as raw material for the production of carrageenans. These components seem to be very useful and promising materials with a number of biological activities. Carrageenans are used in food, cosmetics, and pharmaceuticals. The functionality of carrageenans in various applications depends largely on their rheological properties.

REFERENCES

Barabanova, A. O., & Yermak, I. M. (2012). Structural peculiarities of sulfated polysaccharides from red algae Tichocarpus crinitus (Tichocarpaceae) and Chondrus pinnulatus (Gigartinaceae) collected at the Russian Pacific coast. In *Handbook of marine macroalgae: Biotechnology and applied phycology* (pp. 193–204). West Sussex, UK: John Wiley & Sons.

Blanque, R., Meakin, C., Millet, S., & Gardner, C. (1998). Selective enhancement of LPS-induced serum TNF-α production by carrageenan pretreatment in mice. *General Pharmacology: The Vascular System, 31*, 301–306.

Bornhöft, M., Thommes, M., & Kleinebudde, P. (2005). Preliminary assessment of carrageenan as excipient for extrusion/spheronisation. *European Journal of Pharmaceutics and Biopharmaceutics, 59*, 127–131.

Briones, A., Ambal, W., Estrella, R., Lanto, E., Sison, F., & Villanueva, M. (2000). Anti-blood coagulant activity and hypocholesterolemic property of Philippine carrageenan. *Philippine Journal of Science, 129*, 85–91.

Buck, C. B., Thompson, C. D., Roberts, J. N., Müller, M., Lowy, D. R., & Schiller, J. T. (2006). Carrageenan is a potent inhibitor of papillomavirus infection. *PLoS Pathogens, 2*, e69.

Carlucci, M., Scolaro, L., Noseda, M., Cerezo, A., & Damonte, E. (2004). Protective effect of a natural carrageenan on genital herpes simplex virus infection in mice. *Antiviral Research, 64*, 137–141.

de Souza, M. C. R., Marques, C. T., Dore, C. M. G., da Silva, F. R. F., Rocha, H. A. O., & Leite, E. L. (2007). Antioxidant activities of sulfated polysaccharides from brown and red seaweeds. *Journal of Applied Phycology, 19*, 153–160.

Eccles, R., Meier, C., Jawad, M., Weinmüllner, R., Grassauer, A., & Prieschl-Grassauer, E. (2010). Efficacy and safety of an antiviral iota-carrageenan nasal spray: A randomized, double-blind, placebo-controlled exploratory study in volunteers with early symptoms of the common cold. *Respiratory Research, 11*, 2–10.

Ellis, A., Keppeler, S., & Jacquier, J. (2009). Responsiveness of κ-carrageenan microgels to cationic surfactants and neutral salts. *Carbohydrate Polymers, 78*, 384–388.

Ermak, I., & Khotimchenko, Y. S. (1998). Physical and chemical properties, applications, and biological activities of carrageenan, a polysaccharide of red algae. *Oceanographic Literature Review, 45*, 207–255.

Guiry, M. (2008). *Seaweed site*. Galway: World-Wide Electronic Publication, National University of Ireland.

Haijin, M., Xiaolu, J., & Huashi, G. (2003). A κ-carrageenan derived oligosaccharide prepared by enzymatic degradation containing anti-tumor activity. *Journal of Applied Phycology, 15*, 297–303.

Keppeler, S., Ellis, A., & Jacquier, J. (2009). Cross-linked carrageenan beads for controlled release delivery systems. *Carbohydrate Polymers, 78*, 973–977.

Kim, S.-K., & Pangestuti, R. (2011). Potential role of marine algae on female health, beauty, and longevity. *Advances in Food and Nutrition Research, 64*, 41–55.

Leibbrandt, A., Meier, C., König-Schuster, M., Weinmüllner, R., Kalthoff, D., Pflugfelder, B., et al. (2010). Iota-carrageenan is a potent inhibitor of influenza A virus infection. *PLoS One, 5*, e14320.

Necas, J., & Bartosikova, L. (2013). Carrageenan: A review. *Veterinary Medicine, 58*, 187–205.

Panlasigui, L. N., Baello, O. Q., Dimatangal, J. M., & Dumelod, B. D. (2003). Blood cholesterol and lipid-lowering effects of carrageenan on human volunteers. *Asia Pacific Journal of Clinical Nutrition, 12*, 209–214.

Pujol, C. A., Carlucci, M. J., Matulewicz, M. C., & Damonte, E. B. (2007). Natural sulfated polysaccharides for the prevention and control of viral infections. In *Bioactive Heterocycles V* (pp. 259–281). Germany: Springer-Verlag Berlin Heidelberg.

Pujol, C., Scolaro, L., Ciancia, M., Matulewicz, M., Cerezo, A., & Damonte, E. (2006). Antiviral activity of a carrageenan from Gigartina skottsbergii against intraperitoneal murine herpes simplex virus infection. *Planta Medica*, *72*, 121–125.

Sokolova, E., Barabanova, A., Homenko, V., Solov'eva, T., Bogdanovich, R., & Yermak, I. (2011). In vitro and ex vivo studies of antioxidant activity of carrageenans, sulfated polysaccharides from red algae. *Bulletin of Experimental Biology and Medicine*, *150*, 426–428.

Sun, T., Tao, H., Xie, J., Zhang, S., & Xu, X. (2010). Degradation and antioxidant activity of κ-carrageenans. *Journal of Applied Polymer Science*, *117*, 194–199.

Takeyama, T., Suzuki, I., Ohno, N., Oikawa, S., Sato, K., Ohsawa, M., et al. (1987). Host-mediated antitumor effect of grifolan NMF-5N, a polysaccharide obtained from Grifola frondosa. *Journal of Pharmacobio-Dynamics*, *10*, 644.

Thommes, M., & Kleinebudde, P. (2006). Use of κ-carrageenan as alternative pelletisation aid to microcrystalline cellulose in extrusion/spheronisation. I. Influence of type and fraction of filler. *European Journal of Pharmaceutics and Biopharmaceutics*, *63*, 59–67.

Torsdottir, I., Alpsten, M., Holm, G., Sandberg, A.-S., & Tölli, J. (1991). A small dose of soluble alginate-fiber affects postprandial glycemia and gastric emptying in humans with diabetes. *Journal of Nutrition*, *121*, 795–799.

Tsen, J.-H., Lin, Y.-P., & An-Erl King, V. (2004). Fermentation of banana media by using κ-carrageenan immobilized Lactobacillus acidophilus. *International Journal of Food Microbiology*, *91*, 215–220.

Wijesekara, I., Pangestuti, R., & Kim, S. K. (2010). Biological activities and potential health benefits of sulfated polysaccharides derived from marine algae. *Carbohydrate Polymers*, *81*, 14–21.

Yermak, I. M., & Khotimchenko, Y. S. (2003). Chemical properties, biological activities and applications of carrageenan from red algae. In M. Fingerman, & R. Nagabhushanam (Eds.), *Recent advances in marine biotechnology: Vol. 3.* (pp. 207–255). New York: Science Publisher Inc.

Yuan, Y. V. (2007). Marine algal constituents. In C. Barrow, & F. Shahidi (Eds.), *Marine nutraceuticals and functional foods* (p. 259). Boca Raton, FL: CRC-Taylor and Francis.

Yuan, H., Song, J., Li, X., Li, N., & Liu, S. (2011). Enhanced immunostimulatory and anti-tumor activity of different derivatives of κ-carrageenan oligosaccharides from Kappaphycus striatum. *Journal of Applied Phycology*, *23*, 59–65.

Yuan, H., Song, J., Zhang, W., Li, X., Li, N., & Gao, X. (2006). Antioxidant activity and cytoprotective effect of κ-carrageenan oligosaccharides and their different derivatives. *Bioorganic and Medicinal Chemistry Letters*, *16*, 1329–1334.

Yuan, H., Zhang, W., Li, X., Lü, X., Li, N., Gao, X., et al. (2005). Preparation and in vitro antioxidant activity of κ-carrageenan oligosaccharides and their oversulfated, acetylated, and phosphorylated derivatives. *Carbohydrate Research*, *340*, 685–692.

Zhou, G., Sheng, W., Yao, W., & Wang, C. (2006). Effect of low molecular λ-carrageenan from Chondrus ocellatus on antitumor H-22 activity of 5-Fu. *Pharmacological Research*, *53*, 129–134.

Zhou, G., Sun, Y., Xin, H., Zhang, Y., Li, Z., & Xu, Z. (2004). In vivo antitumor and immunomodulation activities of different molecular weight lambda-carrageenans from Chondrus ocellatus. *Pharmacological Research*, *50*, 47–53.

Zhou, G., Xin, H., Sheng, W., Sun, Y., Li, Z., & Xu, Z. (2005). In vivo growth-inhibition of S180 tumor by mixture of 5-Fu and low molecular λ-carrageenan from Chondrus ocellatus. *Pharmacological Research*, *51*, 153–157.

CHAPTER EIGHT

Biological Activities of Heparan Sulfate

Muthuvel Arumugam[1], Sadhasivam Giji

Centre of Advanced Study in Marine Biology, Faculty of Marine Sciences, Annamalai University, Parangipettai, Tamil Nadu, India
[1]Corresponding author: e-mail address: mamnplab@gmail.com

Contents

1. Introduction 125
2. Materials and Methods 127
   2.1 Isolation 127
   2.2 Anticoagulant activity 128
   2.3 Antiproliferative activity 128
   2.4 ^{1}H-NMR 128
3. Results 129
4. Discussion 130
Acknowledgments 134
References 134

Abstract

Heparan sulfate was isolated from two bivalve mollusks such as *Tridacna maxima* and *Perna viridis*. The isolated heparin was quantified in crude as well as purified samples and they were estimated as 2.72 and 2.2 g/kg (crude) and 260 and 248 mg/g (purified) in *T. maxima* and *P. viridis*, respectively. Both the bivalves showed the anticoagulant activity of the crude and purified sample as 20,128 USP units/kg and 7.4 USP units/mg, 39,000 USP units/kg and 75 USP units/mg, 9460 USP units/kg and 4.3 USP units/mg, and 13,392 USP units/kg and 54 USP units/mg correspondingly in *T. maxima* and *P. viridis*. The antiproliferative activity that was studied with pulmonary artery smooth muscle cells using RPMI media reported that the result is in a dose-dependent manner. Among the two clams, *P. viridis* showed more antiproliferative activity than that of *T. maxima*.

1. INTRODUCTION

Heparan sulfate (HS) has been isolated from tissues of a large number of vertebrate and invertebrate organisms. Invertebrates were first shown to contain a heparin or HS (Burson et al., 1956). An exhaustive assessment showed that mollusks are particularly rich source of the sulfated polysaccharides (Nader & Dietrich, 1989) and it amounts to 90% of the total GAGs content of the mollusks. But the heparin/HS isolated from mollusks are structurally different from human heparin and pharmaceutical heparins (Loganathan, Wang, Mallis, & Linhardt, 1990). Heparin and HS have been the subject of intensive study because of their well-recognized ability to bind proteins that regulate a variety of important biological processes. Heparin and HS are comprised of alternating $1 \rightarrow 4$-linked glucosamine and uronic acid residues. HS is composed of primary monosulfated disaccharides of N-acetyl-D-glucosamine and D-glucuronic acid, while heparin is composed of mainly trisulfated disaccharides of N-sulfoyl-D-glucosamine and L-iduronic acid (Linhardt & Toida, 1996). Recently reported, the diverse biological functions present considerable opportunities for exploiting HS or HS-protein conjugates for developing new classes of anticancer (Shriver, Raguram, & Sasisekharan, 2004), antiviral (Baleux et al., 2009), and improved anticoagulant drugs (Chen, Jones, & Liu, 2007). A method to construct HS is, therefore, critical for developing HS-based therapeutics. Total chemical synthesis is fully capable of preparing short fragments of HS.

HS is found on the surface of endothelial cells. This suggests a physiologically relevant role in regulating levels of anticoagulant activity within the blood vessel. Pharmaceutical heparin is a product of mast cells that is isolated from porcine intestinal tissue with a stronger anticoagulant activity than endothelial HS. Heparin is considered a specialized form of HS with higher levels of sulfation per saccharide unit and iduronic acid content. For example, HS contains about 0.6 sulfo groups per disaccharide unit, while heparin contains 2.6 sulfo groups. Further, about 40% of uronic acid in HS is iduronic, while 90% uronic acid in heparin is iduronic. It plays important roles in regulating numerous functions of the blood vessel wall, including blood coagulation, inflammation response, and cell differentiation. HS is a highly sulfated polysaccharide containing glucosamine and glucuronic/iduronic acid repeating disaccharide units. The unique sulfated saccharide sequences of HS determine its specific functions. Heparin, an analogue of HS, is the most commonly used anticoagulant drug. Because of its wide range of biological functions, HS has become an interesting molecule to biochemists, medicinal chemists, and developmental biologists (Fig. 8.1).

Glycosaminoglycans (GAGs) have been isolated from various tissues obtained from a large number of animal species including both vertebrates and invertebrates. Invertebrates were first shown to contain a heparin or HS (Burson et al., 1956). An exhaustive assessment showed that the mollusks are particularly rich source of these sulfated polysaccharides and it often

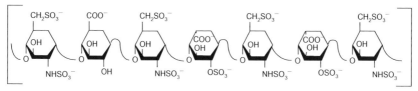

Figure 8.1 The structure showing the repeating disaccharide units.

corresponds up to 90% of the total GAG content these organisms (Nader & Dietrich, 1989). Heparin and heparin-like substances have a wide range of important biological activities including inhibition of pulmonary artery smooth muscle cell (PASMC) proliferation. In the normal physiological state, the smooth muscle cells (SMCs) are entering into quiescent growth state in pulmonary arterial walls which is regulated by a balance between inhibitory and mitogenic factors (Garg, Thomson, & Hales, 2000). The major effect of heparin on blood coagulation is to accelerate the normal rate at which antithrombin III neutralizes the proteolytic activities of several serine proteases in the coagulation sequence. The search for the new sources of heparin with low toxicity prompted many scientists to focus their attention toward marine animals. Earlier studies have shown that heparin-like compounds are also present in some invertebrates (Cassaro & Dietrich, 1977; Patel, Ehrlich, Stivala, & Singh, 1980). A substance denoted "mactin" with anticoagulant activity and structural similarities to mammalian heparins was isolated from the mollusks *Cyprinia islandica* and *Mactrus pussula*. Another compound from the clam *Mercenaria mercenaria* also exhibited several structural similarities to heparin. Previously conducted detailed investigations on the heparin isolated from the clams (Pejler, Danielsson, Bjork, & Lindahl, 1987) *A. brasiliana* and *T. mactroides* found that the clam heparin preparations have basically the same structure as mammalian heparins, but notable differences included the greater chain length of clam heparin. Further heparin from some mollusks was reported to show difference in their anticoagulant activity (Dietrich et al., 1989). The present study aimed to find out the source of heparin and discuss their antiproliferative and anticoagulant activity.

This chapter reports the isolation of GAGs and their structural features with special reference to heparin/HS from marine bivalve mollusks.

2. MATERIALS AND METHODS

2.1. Isolation

The standard procedure was followed for the extraction of heparin/HS, with suitable modification (Arumugam & Shanmugam, 2004) for the defating and deproteinization of *T. maxima* and *P. viridis* treated with CPC. The purification of the crude HS was done by the using the anionic resin (Amberlite IRA-900 Cl⁻). The purified GAGs were converted into heparin sodium/HS salts by using cationic resin (Amberlite IR-120 Na⁺) and the recovered precipitate was taken for further analyses (Nishino, Yokoyama, Dobashi, Fujihara, & Nagumo, 1989).

2.2. Anticoagulant activity

Heparin readily catalyzes the inactivation of factor Xa by anti-thrombin III. Factor Xa inactivation was used in this study to assess the anticoagulant activity of the GAGs prepared from both mussels using a Heparin Assay Kit (Sigma). In this assay, when both factor Xa and anti-thrombin III are present in excess, the inhibition of factor Xa is directly proportional to the limiting concentration of heparin. Thus, residual factor Xa activity, measured with factor Xa-specific chromogenic substrate, is inversely proportional to the heparin concentration (Teien, Lie, & Abilguard, 1976).

2.3. Antiproliferative activity

The antiproliferative activity of samples was determined by the MTT assay (Mosmann, 1983). The isolated bovine PASMCs were seeded at 1.5×10^4 cells/well into a 6-well tissue culture plate, grown for 2 days; then growth was arrested at the end of 48 h by reducing the serum concentration of the medium from 0.1% to 10%. The media was then changed to the experimental samples which contained either standard media (RPMI-1640 with 10% fetal bovine serum (FBS) (Sigma, St. Louis, Mo)), growth arrest media (RPMI with 0.1% FBS), or standard media with oligomers/heparin (5 μg/ml). All media contained streptomycin (100 μg/ml), Penicillin (100 U/ml), and amphotericin B (1.25 μg/ml). After 4–5 days of growth, the cells were lifted with trypsin/EDTA and then counted using a Coulter Counter. Results are presented as mean ± standard error of the mean. Comparisons among groups were made with a factorial analysis of variance (ANOVA), using the STATE VIEW software package (Brainpower, Inc., to Calabasas, CA) for Macintosh computers.

2.4. ¹H-NMR

NMR spectroscopy was analyzed on samples dissolved in D_2O to remove exchangeable protons and transferred to Shigemi tubes. ¹H-NMR experiments were performed on an AMX-400 at 400 MHZ.

3. RESULTS

The amount of HS was estimated as 2.72 g/kg of dry tissue in *T. maxima* and 2.2 g/kg in *P. viridis* (crude). After purification by using the Amberlite anion exchange resin, the yield was found to be 260 and 248 mg/g, respectively, in *T. maxima* and *P. viridis*, as presented in Table 8.1.

In the anticoagulant assay (Table 8.2), the yield and anticoagulant activity of the *T. maxima* crude sample were reported to be 20,128 USP units/kg and 7.4 USP units/mg; whereas the purified sample showed 39,000 USP units/kg and 75 USP units/mg of the sample, respectively. Likewise, in the case of *P. viridis*, the yield and anticoagulant activity were estimated as 9460 USP units/kg and 4.3 USP units/mg in the crude sample and 13,392 USP units/kg and 54 USP units/mg in the purified sample, respectively.

The effect of HS (GAGs) isolated from *P. viridis* and *T. maxima* and Upjohn heparin grown in RPMI medium containing 0.1% and 10% FBS is depicted in Fig. 8.2. The heparin/HS isolated from *P. viridis* recorded increasing inhibition over the growing cells when the concentration increased from 1.0 to 100 μg, i.e., the percentage cell mean in the case of 1.0 μg was found to be 61.764±3.660; in 10 μg, it was 33.064±3.507;

Table 8.1 The yield of crude and purified heparan sulfate complex of *T. maxima* and *P. viridis*

S. no.	Source	Net yield	
		Crude (g/kg)	Purified (mg/g)
1.	*Tridacna maxima*	2.72	260
2.	*Perna viridis*	2.2	248

Table 8.2 The anticoagulant activities of the heparan sulfate extracted from *T. maxima* and *P. viridis*

S. no.	Origin	Anticoagulant assay (USP method)			
		Crude		Purified	
		Activity (IU/mg)	Yield (IU/kg)	Activity USP (units/mg)	Yield USP (units/kg)
1.	*Tridacna maxima*	7.4	20,128	75	39,000
2.	*Perna viridis*	4.3	9460	54	13,392

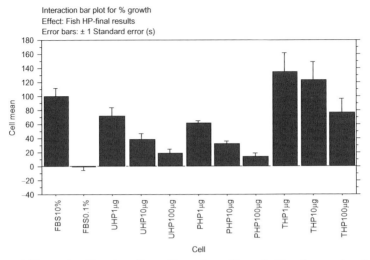

Figure 8.2 The antiproliferative effect of heparin and heparin like glycosaminoglycans from *P. virdis* (PHP) & *T. maxima* (THP) and Upjohn heparin.

and in 100 μg concentration, it was 15.071 ± 4.609, showing an increasing influence over the cell growth. The inhibition of *P. viridis* is comparatively more even than that of Upjohn heparin, the standard used in the present study, which showed the percentage mean cell growth as follows: 1 μg—72.533 ± 11.200, 10 μg—8.809 ± 7.940, and 100 μg—19.168 ± 5.921, respectively.

But on the contrary, the heparin and heparin-like GAGs extracted from *T. maxima* were found to promote the cell growth in lower concentrations, i.e., 1 μg—134.939 ± 26.468% and 10 μg—123.934 ± 25.325%. But at the same time, at higher concentration it was also found to be reducing the growth (at 100 μg concentration, the mean cell growth was only 77.429 ± 18.923%).

^{1}H-NMR analysis revealed the presence of signals corresponding to GlcNac (*N*-acetyl methyl at 1.99 ppm and H-1 at 5.4 ppm) in both the samples (Figs. 8.3 and 8.4).

4. DISCUSSION

In the present investigation, the yield of crude HS was found to be 2.72 g/kg and 2.2 g/kg in *T. maxima* and *P. viridis*, respectively, whereas the HS yield was 2.2–2.8 g/kg, 1.8–2.5 g/kg, and 2.7–3.8 g/kg in *A. brasiliana*, *D. striatus*, and *T. mactroides*, respectively (Dietrich et al.,

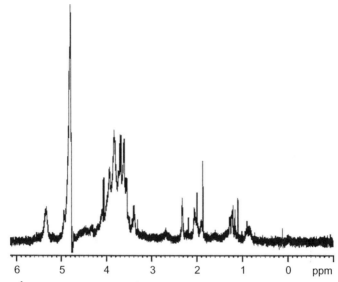

Figure 8.3 ¹H-NMR spectroscopy of heparan sulfate from *Tridacna maxima*.

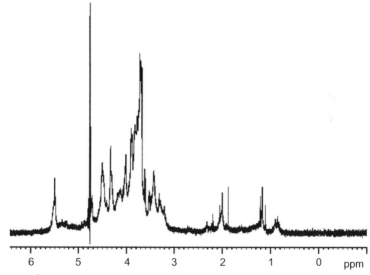

Figure 8.4 ¹H-NMR spectroscopy of heparan sulfate from *Perna viridis*.

1989). At the same time, the yield of heparin and other sulfated mucopolysaccharides from thymus is 274 µg/kg only (Straus, Nader, Bianchini, & Osima, 1981). Likewise, the sulfated mucopolysaccharides were isolated by using quaternary ammonium salts (Barlow, Coen, & Mozen, 1964) and the yield was reported to be 170, 174, 843, 307, and 1090 µg/kg dry tissue in different molluscan species such as *Aulocombia ater*, *Perna perna*, *Mesodesma donacium*, *Loligo brasiliense*, and *Octopus* species, respectively.

In the present study, the crude sample of *T. maxima* and *P. viridis* showed 7.4 USP units/mg and 4.3 USP units/mg of anticoagulant activity, respectively. But in the purified sample of *T. maxima* and *P. viridis*, the anticoagulant activity was found to be higher (75 USP units/mg and 54 USP units/mg, respectively) than that of the crude sample which is further higher than that of *K. opima* (39.7 BP units/mg), sheep heparin (59 USP/mg), fin back whale *Balaenoptera physalus* (40–70 USP units/mg), *H. pugilinus* (26 USP units/mg) (Barlow et al., 1964), and HS anticoagulant activity (Nader et al., 1983) in *Pomacea* sp., *T. gibbus*, *A. brasiliana*, beef pancreas (<5 IU/mg), and lobster HS (16 IU/mg), respectively (Hovingh & Linker, 1982).

The anticoagulant assay in the purified sample of *T. maxima* (39,000 USP units/kg) is lower than that of the *K. opima* (Somasundaram, 1990) (283,370 USP units/kg), and *A. rhombea* (74,140 USP units/kg). But at the same time, it is higher than that of *C. madrasensis* (Hovingh & Linker, 1982) (14,125 USP units/kg) and *O. vulgaris* (Somasundaram, 1990) (12,960 USP units/kg). Whereas in the case of *P. viridis*, the yield in the anticoagulant assay was reported as 9460 USP units/kg and 13,392 USP units/kg in crude and purified sample, respectively. This variation might be due to the presence of nonanticoagulant substance in the samples since the activity of heparin depends upon the amount of impurity carried over in the isolated products (Berteau, Mc Cort, Goasdou, Tissot, & Daniel, 2002).

Antiproliferative activity of isolated heparin/HS from *P. viridis* could also be compared with that of Dahlberg et al. (1996) in which they reported such inhibition over the cell growth in Elkins–Sinn heparin on PASM cells. But at the same time, they also observed a stimulatory effect on PASM cell growth (1.4% increase) at 1.0 µg/ml concentration of Choay heparin which has also reported less antiproliferative activity of $29 \pm 5\%$ at 10 µg/ml concentration like the heparin GAGs extracted from *T. maxima*.

Further it is interesting to note that the same commercial standard Upjohn heparin showed varying antiproliferative activity on the PASM cells in a medium containing 0.1% FBS. In the present study, it produced 27.36% and 61.19% of inhibition on the PASM cells grown in RPMI medium containing 0.1% and 10% FBS; whereas it showed an antiproliferative activity of 48 ± 5 and 80 ± 4 at the concentration of 1 or 10 µg/ml (Dahlberg et al., 1996).

The results reveal that there is a dose-dependant decrease in the percentage of viable cells when treated with the heparin/HS from *T. maxima* and *P. viridis* and standard heparin. There was a significant reduction in the number of live cells in the case of all the concentrations of heparin GAGs from *P. viridis* in an ascending order with increasing concentration. But, at the same time, the isolated compound from *T. maxima* did not show any inhibitory effect at 1 and 10 µg concentrations and it showed some inhibitory effect at 10 µg concentration. From the above results, it is quite evident that the heparin GAGs isolated from *P. viridis* has more antiproliferative effect than even the commercial standard Upjohn heparin and that of *T. maxima*.

Apart from the above, the previous investigations have shown variable effects of heparin on cell growth. Heparin did not inhibit endothelial cell proliferation (Dahlberg et al., 1996; Zaragosa, Battle-Tracy, & Owen, 1990) whereas demonstrated the inhibition over the aortic SMC growth. When using the rat aortic SMCs, reported varying antiproliferative activity in 10 different commercial heparins (Zaragosa et al., 1990). Among them, two lots of heparin stimulated the cell growth at low concentration as the HS isolated from *T. maxima*. Therefore, both the cell type to be studied and the source of heparin are important in determining the antiproliferative activity of heparins. Further, for as yet uncertain chemical reasons, not all lots of heparin have the antiproliferative activity to the same degree (Dahlberg et al., 1996). So the foregoing account may shed light on the previous paradoxes in which different commercial heparin/HS GAGs isolated from different sources have had variable antiproliferative effects even on the same cell types.

Further the inhibiting effect of heparin on SMC growth *in vitro* has been shown to be independent of its anticoagulant activity (Griffin et al., 1995; Tiozzo et al., 1991). This concept is very well supported by the results of the present study also. In the present investigation, though the purified heparin/HS extracted from *T. maxima* reported more anticoagulant activity (30,212 IU/kg) than that of *P. viridis* (16,144 IU/kg), the heparin/HS from *P. viridis* showed more antiproliferative activity than that of *T. maxima* which showed stimulatory effect also in lower concentrations (1 and 10 µg) studied. This suggests that the anticoagulant activity does not correlate with the ability of a given heparin GAGs to inhibit the cell growth. From the foregoing account, it could be concluded that the heparin/HS isolated from *P. viridis* and *T. maxima* has an anti-cell proliferative component which needs to be further studied for using it as an anticancer drug.

The precise relationship between the structure of the mucopolysaccharide and its biologic properties has remained elusive. Hence, analysis of the structure is essential in natural products for correlation of structure–function relationship. In conclusion, the proton NMR spectra of isolated HS are fully compatible in their characteristics with that of standard heparin.

ACKNOWLEDGMENTS
The authors are thankful to the authorities of Annamalai University for providing the facilities. We are thankful to Dr. Bruna Parma, Opocrin Research Institute, Brazil, for useful suggestions in designing this work.

REFERENCES
Arumugam, M., & Shanmugam, A. (2004). Extraction of heparin and heparin-like substances from marine mesogastropod mollusc *Turritella attenuata* (Lamarck, 1779). *Indian Journal of Experimental Biology*, 42, 529–532.
Baleux, F., Loureiro-Morais, L., Hersant, Y., Clayette, P., Arenzana-Seisdedos, F., Bonnaffe, D., et al. (2009). Synthetic CD4-heparan sulfate glycoconjugate inhibits CCR5 and CXCR4 HIV-1 attachment and entry. *Nature Chemical Biology*, 5, 743–748.
Barlow, G. H., Coen, L. J., & Mozen, M. M. (1964). A biological, chemical and physical comparison of heparin from different mammalian species. *Biochimica et Biophysica Acta*, 83, 272–277.
Berteau, O., Mc Cort, I., Goasdou, N., Tissot, B., & Daniel, R. (2002). Characterization of a new α-L- fucosidase isolated from the marine mollusk *Pecten maximus* that catalyses the hydrolyses of α-L- fucose from algal fucoidan (*Ascophyllum nodosum*). *Glycobiology*, 12, 273–282.
Burson, S. L., Fahrenbach, M. J., Fommhagen, L. H., Riccardi, B. A., Brown, R. A., Brockman, J. A., et al. (1956). Isolation and purification of mactins, heparin-like anticoagulants from mollusca. *Journal of the American Chemical Society*, 78, 5874–5879.
Cassaro, C. M. F., & Dietrich, C. P. (1977). The distribution of sulfated mucopolysaccharides in invertebrates. *Journal of Biological Chemistry*, 252, 2254–2261.
Chen, J., Jones, C. L., & Liu, J. (2007). Using an enzymatic combinatorial approach to identify anticoagulant heparan sulfate structures. *Chemistry and Biology*, 14, 986–993.
Dahlberg, C. G. W., Thompson, B. T., Joseph, P. M., Garg, H. G., Spence, C. R., Quinn, D. A., et al. (1996). Differential effect of three commercial heparins on Na^+/H^+ exchange and growth of PASMC. *American Journal of Physiology—Lung Cellular and Molecular Physiology*, 270, L260–L265.
Dietrich, C. P., Nader, H. B., Paiva, J. F., Santos, E. A., Holme, K. R., & Perlin, A. S. (1989). Heparin in molluscs: Chemical enzymatic degradation and super (13) C and super (1) H NMR spectroscopial evidence for the maintenance of the structure through evolution. *International Journal of Biological Macromolecules*, 11, 361–366.
Garg, H. G., Thomson, T., & Hales, C. A. (2000). Structural determinants of antiproliferative activity of heparin on pulmonary smooth muscle cells. *American Journal of Physiology Lung Cellular and Molecular Physiology*, 279, L779–L789.
Griffin, C. C., Linhardt, R. J., Van Gorp, C. L., Toida, T., Hileman, R. E., Schubert, R. C., et al. (1995). Isolation and characterization of heparan sulfate from crude porcine intestinal mucosa peptidoglycan heparin. *Carbohydrate Research*, 276, 183–197.
Hovingh, P., & Linker, A. (1982). An unusual heparan sulfate isolated from lobsters (*Homarus americanus*). *Journal of Biological Chemistry*, 257, 9840–9844.
Linhardt, R. J., & T.Toida, 1996. *Heparin oligosaccharide-new analogs: Development and application, in carbohydrates as Drugs* (pp. 277–341). (Z. B. Wite Zale & K. A. Nie forth, Eds.). New York, NY: Marker Dekker.

Loganathan, D., Wang, H. M., Mallis, L. M., & Linhardt, R. J. (1990). Structural variation in the antithrombin III binding site region and its occurence in heparin from different sources. *Biochemistry, 29*, 4362–4368.

Mosmann, T. (1983). Rapid colorimetric assay for cellular growth and survival: Application to proliferation and cytotoxicity assays. *Journal of Immunological Methods, 65*, 55–63.

Nader, H. B., & Dietrich, C. P. (1989). Natural occurrence possible biological role of heparin. In D. A. Lane, & U. Lindhal (Eds.), *Heparin: Chemical and biological properties. Clinical application* (p. 81.3). London/Melbourne/Auckland: Edward Arnold.

Nader, H. B., Medeiros, M. G. L., Paiva, J. F., Paiva, V. M. P., Jeronino, S. M. B., Ferreira, T. M. P. C., et al. (1983). A correlation between the sulfated glycosaminoglycan concentration and degree of salinity of the habitat in fifteen species of the classes Crustacea, Pelecypoda and Gastropoda. *Comparative Biochemistry and Physiology, 76*, 433–436.

Nishino, T., Yokoyama, G., Dobashi, K., Fujihara, M., & Nagumo, T. (1989). Isolation, purification and characterization of fucose containing sulfated polysaccharides from the brown sea weed Eclonia kurome and their blood anticoagulant activities. *Carbohydrate Research, 186*(1), 119–129.

Patel, B., Ehrlich, J., Stivala, S. S., & Singh, N. K. (1980). Comparative studies of mucopolysaccharides from marine animals *Raja eglanteria* Bosc. *Journal of Experimental Marine Biology and Ecology, 46*, 127–136.

Pejler, G., Danielsson, A., Bjork, I., & Lindahl, V. (1987). Structure and antithrombin binding properties of heparin isolated from the clams *A. brasiliana* and *T. mactroides*. *Journal of Biological Chemistry, 262*, 11413–11421.

Shriver, Z., Raguram, S., & Sasisekharan, R. (2004). Glycomics: A pathway to a class of new and improved therapeutics. *Nature Reviews Drug Discovery, 3*, 863–873.

Somasundaram, S. T. (1990). *Heparin like (glycosaminoglycan) from Katelysia opima*. Ph.D. thesis, (p. 54). Portonovo, India: Annamalai University.

Straus, A. H., Nader, H. B., Bianchini, P., & Osima, B. (1981). Isolation and pharmacological activities of heparin and other sulfated mucopolysaccharides from Thymus. *Biochemical Pharmacology, 30*(10), 1077–1981.

Teien, A. N., Lie, M., & Abilguard, U. (1976). Assay of heparin in plasma using a chromogenic substrate for activated factor Xa. *Thrombosis Research, 8*, 413–416.

Tiozzo, R., Reggiani, D., Cingi, M. R., Bianchini, P., Osima, B., & Calandra, S. (1991). Effect of heparin derived fractions on the proliferation and protein synthesis of cells in culture. *Thrombosis Research, 62*, 177–188.

Zaragosa, R., Battle-Tracy, K. M., & Owen, N. E. (1990). Heparin inhibits Na^+/H^+ exchange in vascular smooth muscle cells. *American Journal of Physiology, 258*(27), C46–C53.

CHAPTER NINE

Beneficial Effects of Hyaluronic Acid

Prasad N. Sudha*,[1], Maximas H. Rose[†]

*PG and Research Department of Chemistry, DKM College for Women, Thiruvalluvar University, Vellore, Tamil Nadu, India
[†]Department of Biology, Sri Sai Vidyasharam, Vellore, Tamil Nadu, India
[1]Corresponding author: e-mail address: drparsu8@gmail.com

Contents

1. Introduction — 138
2. Structure of Hyaluronic Acid — 139
3. Properties of Hyaluronic Acid — 141
4. Modification of Hyaluronic Acid — 141
5. Applications of Hyaluronic Acid — 146
   5.1 Biomedical applications — 146
   5.2 TE applications — 152
   5.3 Drug delivery applications — 160
   5.4 Gene delivery applications — 162
   5.5 Targeted drug delivery — 163
   5.6 HA hydrogels — 165
   5.7 Tumor treatment — 166
   5.8 Environmental applications — 167
   5.9 Sensors — 168
6. Conclusion — 168
Acknowledgments — 168
References — 168

Abstract

Biomaterials are playing a vital role in our day-to-day life. Hyaluronan (hyaluronic acid), a biomaterial, receives special attention among them. Hyaluronic acid (HA) is a polyanionic natural polymer occurring as linear polysaccharide composed of glucuronic acid and N-acetylglucosamine repeats via a β-1,4 linkage. It is the most versatile macromolecule present in the connective tissues of all vertebrates. Hyaluronic acid has a wide range of applications with its excellent physicochemical properties such as biodegradability, biocompatibility, nontoxicity, and nonimmunogenicity and serves as an excellent tool in biomedical applications such as osteoarthritis surgery, ocular surgery, plastic surgery, tissue engineering, and drug delivery. It plays a key role in cushioning and lubricating the body and is abundant in the eyes, joints, and heart valves. A powerful antioxidant, hyaluronic acid is perhaps best known for its ability to bond water to tissue. Hyaluronan production increases in proliferating cells, and the polymer may play a role in mitosis.

This chapter gives an overview of hyaluronic acid and its physicochemical properties and applications. This chapter gives a deep understanding on the special benefits of hyaluronic acid in the fields of pharmaceutical, medical, and environmental applications. Hyaluronic acid paves the way for beneficial research and applications to the welfare of life forms.

1. INTRODUCTION

In 1934, Karl Meyer and his assistant, John Palmer, described a procedure for isolating a novel glycosaminoglycan (GAG) from the vitreous of bovine eyes. They showed that this substance contained an uronic acid and an amino sugar but without sulfoesters. In their words: "we propose, for convenience, the name 'hyaluronic acid', from hyaloid (vitreous) + uronic acid." This marked the birth announcement for one of nature's most versatile and fascinating macromolecules "hyaluronan," as is most frequently referred to. Hyaluronan is present in all vertebrates as an essential constituent of the extracellular matrices (ECMs) in most mature tissues. It is a major constituent in the vitreous humor of the eye (0.1–0.4 mg/g wet weight in human), in synovial joint fluid (3–4 mg/ml), in the matrix produced by the cumulus cells around the oocyte prior to ovulation (∼0.5 mg/ml), in the pathological matrix that occludes the artery in coronary restenosis, etc. (Hascall, Calabro, Oken, & Masellis, 2002).

The largest quantity of hyaluronan (7–8 g per average adult human, ∼50% of the total in the body) resides in skin tissue, where it is in both the dermis (∼0.5 mg/g wet tissue) and the epidermis (∼0.1 mg/g wet tissue). Interestingly, dermis consists primarily of ECM with a sparse population of cells, while the epidermis is in the reverse; the keratinocytes fill all but a few percent of the tissue. Thus, the actual concentration of hyaluronan in the matrix around the cells in the epidermis (estimated to be 2–4 mg/ml) is an order of magnitude higher than in the dermis (estimated to be ∼0.5 mg/ml) (Fraser, Laurent, & Laurent, 1997).

The concentration of hyaluronan in tissues is often higher than would be expected if individual molecules maintained their expanded domain structures. In many cases, the hyaluronan is organized into the ECM by specific interactions with other matrix macromolecules. However, high-molecular-weight (HMW) hyaluronan at high concentration in solution (e.g., 5 MDa at concentrations above 0.1 mg/ml) can also form entangled molecular networks through steric interactions and self-association between and within individual molecules. The latter can occur when a stretch of the hydrophobic face of the ribbon structure of the backbone interacts reversibly with the hydrophobic face on a comparable stretch of hyaluronan on another molecule or in a different region of the same molecule. Such networks exhibit different properties than would isolated hyaluronan molecules.

2. STRUCTURE OF HYALURONIC ACID

GAGs are a class of natural macromolecules, which is negatively charged, linear heteropolysaccharides classified into several groups on the basis of structure such as hyaluronic acid, heparin sulfate, dermatan sulfate, and chondroitin sulfate, exhibiting an attracting interest because of their several applications in the biomedical, veterinary, pharmaceutical, and cosmetic field (Linhardt, 2001; Sasisekharan, Raman, & Prabhakar, 2006). Hyaluronic acid (HA) (Fig. 9.1) is a naturally occurring linear polysaccharide composed of glucuronic acid and N-acetylglucosamine repeats via a β-1,4 linkage. It is a linear polymer consists of repeating disaccharide units of N-acetyl-D-glucosamine and D-glucuronic acid linked by β(1,4) and β(1,3) glycosidic linkages, respectively, and molecular mass in the range from 104 to 107 Da. It is distinct from other GAGs by the nonexistence of sulfated groups and lack of covalently linked peptide in their structure (Puré & Assoian, 2009; Stern, 2003).

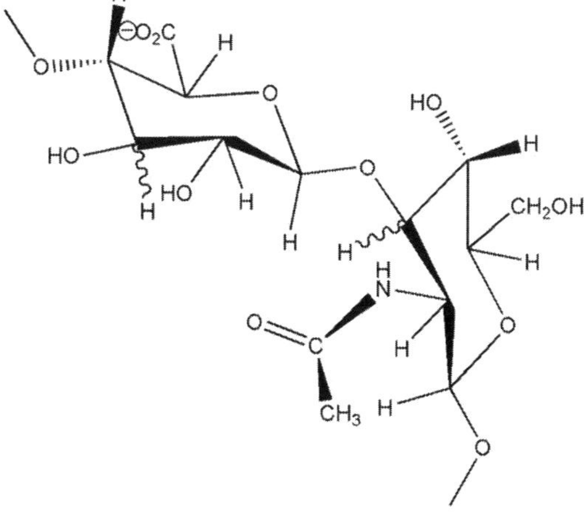

Figure 9.1 Structure of hyaluronic acid (HA).

HA is the most important substance in the synovial fluid (SF) of articular joints, acting as a lubricant for the cartilage and regulating the viscosity of the synovial fluid. Healthy joints contain approximately 2.26 g/L HA (Ziedler, 1986). HA macromolecular chains are built from D-glucuronic acid and N-acetyl-D-glucosamine disaccharides. A molecule of 10 MDa contains 25,000 disaccharide units in the chains, which are held together by hydrophobic bonds (Romagnoli et al., 2008). The polysaccharide chains are linear and unbranched and roll up into a coil conformation. These coils can straighten, and this behavior is the mechanism of action behind viscosupplementation. The length of the polysaccharide chains and the MW of the HA are very different in various tissues. In normal tissues, a molecule of HA (10 MDa) has a thickness of 1 nm and a length of 25 mm (Romagnoli & Belmontesi, 2008). In the biomatrix, HA has an MW in the range of 6–12 MDa (Balazs, 2009). The molecular weight of HA is approximately 7 MDa in healthy joints and 4.8 MDa in unhealthy joints (Wohlrab, Neubert, & Wohlrab, 2004). HA is a polymer formed by repeating disaccharide units of N-acetyl-D-glucosamine and glucuronic acid.

This GAG is present in tissues as the cartilage, synovial fluid, skin, rooster combs, umbilical cord, and vitreous humor and synovial fluid, as well as in the cell wall of bacteria such as *Streptococcus zooepidemicus* (Shiedlin et al., 2004; Vázquez et al., 2009; Yamada & Kawasaki, 2005). HA is abundantly present in almost all biological fluids and tissues. For experimental purposes, commercial HA samples are mostly of bacterial origin (gram-positive *Streptococci*) or isolated from rooster combs, while in the human organism, it is distributed in the skin, vitreous humor of the eye, umbilical cord, cartilage, and synovial fluid. In the latter, methods for HA production have emphasized mainly on the environmental or culture conditions for cell growth and HA formation rather than focusing on cellular metabolism and its regulation for obtaining higher yield and molecular weight (Johns, Tang Goh, & Oeggeru, 1994). The biosynthesis of HA faces a stiff competition between glycolytic pathway and cell wall synthesis (Liu, Du, Chen, Wang, & Sun, 2008). It has also been reported that the molecular weight of HA is controlled by the concentration of the precursor UDP-N-acetylglucosamine, which is limiting compared with the other precursor UDP-glucuronic acid (Chen, Marcellin, Hung, & Nielsen, 2009). But to increase the productivity and the molecular weight of HA, a balanced flux of these precursors toward HA biosynthesis was analyzed by Shah, Badle, and Ramachandran (2013) and found that concentration and molecular weight of HA are increased by decreasing carbon flux toward glycolysis and pentose phosphate pathway and increasing carbon flux toward HA precursor formation. Also, the addition of antioxidant tannic acid also increased molecular weight to 3.0 MDa.

HA serves to maintain its viscoelastic properties required for lubrication of the joint. When hyaluronan is degraded by the action of free radicals or certain enzymes, synovial fluid loses its lubricating properties, which leads to increased wear of the joint and results in arthritic pain. HA has a wide variety of cosmetic and pharmaceutical applications, for instance, to fill soft tissue defects such as facial wrinkles or to treat articular disorders in horses.

3. PROPERTIES OF HYALURONIC ACID

The HA solutions' rheological properties are very important for these applications; for example, ophthalmic viscosurgical devices are classified according their rheological behavior. The rheological characterization of hyaluronic acid aqueous solutions has been carried out by Garcia-Abuin, Gomez-Diaz, Navaza, Regueiro, and Vidal-Tato (2011), determining the value of the intrinsic viscosity and the average molecular weight. The influence of the polymer concentration, temperature, and presence of an electrolyte on the magnitude of the viscosity, density, speed of sound, and rheological behavior has been analyzed. The presence of HA in an aqueous solution showed a complex rheological behavior, including this system into the pseudoplastic fluids. Both the increase of temperature and the presence of an electrolyte produced an important decrease on the viscosity magnitude, as well as an approximation to the Newtonian behavior in relation to its rheology. Procedures for introducing covalent cross-links in hyaluronan matrices have been developed to create stable networks and semisolid materials exhibiting pronounced viscoelastic properties (Laurent, 1998).

4. MODIFICATION OF HYALURONIC ACID

The potential application of HA is limited due to its high water solubility and rapid degradation in the human body. Chemical modification is applied to improve the physicochemical stabilities while retaining its natural biocompatibility, biodegradability, and nonimmunogenicity (Luo, Kirker, & Prestwich, 2000). The general methods to produce HA derivatives include esterification, cross-linking, grafting, and composite modification. Composite modification is attracting more and more attention for its unique advantages compared with other modification methods.

Chemical modifications have been mostly performed on HA through grafting techniques, and, in spite of several works published in this field by employing "grafting to" technique, the control of molecular weight (MW) and polydispersity index (PDI) is still a major issue that needs to be addressed since these two parameters affect the chemical, physical, and biological properties of the final product. Atom transfer radical polymerization (ATRP), discovered by Matyjaszewski and Sawamoto in 1995 (Wang & Matyjaszewski, 1995), is a versatile controlled radical polymerization process. It enables a precise control of MW, PDI, and functionality (Coessens, Pintauer, & Matyjaszewski, 2001).

Hyaluronic acid was also modified by grafting various amino acids on its carboxylic group and then Schanté, Zuber, Herlin, and Vandamme (2012) evaluated the enzymatic stability of the various conjugates in the presence of a hyaluronidase. The results showed that all amino acid–modified HA polymers were more resistant to degradation compared with the native HA albeit with variation according to the amino acids. The findings are consistent with results reported previously (Ibrahim, Kang, & Ramamurthi, 2010) and confirm the beneficial effect of the carboxyl protection of HA with amino acids.

HA was cross-linked with carboxymethylcellulose sodium (CMC-Na), a carbohydrate-derived biomaterial, to form a novel HA-based composites by Liu, Liu, Wang, Du, and Chen (2007). A series of sponge-like composites were prepared by cross-linking different amounts of hyaluronic acid (HA) and CMC-Na. Adipic dihydrazide (ADH) was employed as the cross-linker and water-soluble 1-ethyl-3-[3-(dimethylaminopropyl)] carbodiimide as the carboxyl-activating agent. The prepared composites showed high swelling ratio, improved physicochemical stability, and high antioxidant ability, which makes them potential materials to be used as dermal fillers or applied in soft tissue augmentation or filling.

HMW hyaluronic acid in powder form (HMWHA, with average molecular weight of 1042 kDa) was degraded to low-molecular-weight hyaluronic acid (LMWHA, 200–230 kDa) by several methods, and the changes in molecular structure and antioxidative activities brought about by each degradation method were compared. The degradation methods used were electron beam irradiation (EB), gamma ray irradiation (GM), microwave irradiation, and thermal treatment. The FT-IR spectra showed no substantial changes of the spectral pattern between HMWHA and LMWHA. However, the ultraviolet (UV) absorbance of LMWHA by MW was considerably greater at 265 nm, indicating the formation of more double bonds. The antioxidative activities of all LMWHA samples were found to have risen, but the MW-treated LMWHA showed the most significant increase due to a newly formed double bond. EB- and GM-treated LMWHA showed the lowest polydispersity and little change in UV spectra from those of HMWHA (Choi, Kim, Kim, Kweon, & Lee, 2010).

Recently, LMW hyaluronic acid has been reported to have novel features, such as free radical scavenging activities, antioxidant activities, and promotion of excisional wound healing (Ke, Sun, Qiao, Wang, & Zeng, 2011). It has also been reported that GM of the native hyaluronic acid could increase its antioxidant activity as a result of a decrease in molecular weight (Kim, Ravichandran, Khan, & Kim, 2008). The hyaluronic acids with an average molecular weight in the range of 45.2–145 kDa were shown to possess pronounced free radical scavenging and antioxidant activities, particularly compared with the native hyaluronic acid of 1050 kDa. LMW hyaluronic acid was prepared through degradation of native hyaluronic acid by ozone treatment (Yue, 2012).

In recent years, the deposition of polyelectrolyte multilayers by layer-by-layer (LBL) technique has emerged as a promising tool for the functionalization of various substrates due to their ease of formation and flexibility of tailoring physicochemical properties (Boudou, Crouzier, Ren, Blin, & Picart, 2010). An attempt for the functionalization of photocross-linked HA hydrogels by deposition of poly(L-lysine) (PLL) and HA multilayer films made by the LBL technique was made by Yamanlar, Sant, Boudou, Picart, and Khademhosseini (2011). Modification of HA hydrogel surfaces with multilayer films affected their physicochemical properties and improved cell adhesion and spreading of NIH3T3 fibroblasts cells on these surfaces, making them suitable for various biomedical and tissue engineering (TE) applications including growth factor delivery and coculture systems.

The synthesis and physicochemical characterization of mixed lipoic and formic esters of hyaluronan (Lipohyal) were carried out by Picotti et al. (2013). The synthesis was conducted by activating lipoic acid with 1,1-carbonyldiimidazole to obtain lipoyl imidazolide, which reacted with hyaluronan (HA) in formamide under basic conditions. Lipohyal can be easily cross-linked by UV irradiation, resulting in an innovative hydrogel with distinctive viscoelastic properties that is suitable both as a dermal filler and as an intra-articular medical device.

It has been revealed that coupling of polyelectrolytes with ionic surfactants is a convenient method for the preparation of ionic complexes with remarkable structure and properties (Macknight, Ponomarenko, &

Tirrell, 1998). Specifically, coupling of polyacids with tetraalkylammonium surfactants bearing long alkyl chains is known to lead to amphiphilic comb-like systems displaying a layered biphasic structure able to lodge agents with chemical or biomedical activity. Stoichiometric complexes of hyaluronic acid with alkyltrimethylammonium surfactants bearing octadecyl, eicosyl, and docosyl groups were prepared by ionic coupling in aqueous solution by Tolentino, Alla, de Ilarduya, and Guerra (2013). The complexes were insoluble in water but soluble in organic solvents. In the solid state, they self-assembled in a biphasic layered structure with the alkyl side chains forming a separate phase that melted in the 50–60 °C range. They were stable to heating up to above 200 °C.

An efficient method for the synthesis of hyaluronic acid-based brush copolymers using ATRP has been reported by Pitarresi et al. (2013). At first, two different hyaluronic acid (HA)-based macroinitiators have been prepared, and then they have been used for the polymerization via ATRP of hydrophilic or hydrophobic molecules carrying vinyl portions with two macroinitiators (HA–TBA–BMP and HA–TBA–EDA–BMP). Then they have been used for the ATRP of poly(ethylene glycol) methacrylate (PEGMA), butyl methacrylate (BUTMA), or N-isopropylacrylamide (NIPAM) using a complex of Cu(I) and 2,2′-Bipyridyl (Bpy), as a catalyst. Both macroinitiators and final copolymers, named as HA–BMP–pPEGMA, HA–BMP–pBUTMA, HA–BMP–pNIPAM, HA–EDA–BMP–pPEGMA, HA–EDA–BMP–pBUTMA, and HA–EDA–BMP–pNIPAM, have been characterized by spectroscopic analysis and size exclusion chromatography to confirm the success of the polymerization process.

A series of thermosensitive copolymer hydrogels, aminated hyaluronic acid-g-poly(NIPAM) (AHA-g-PNIPAAm), were synthesized by coupling carboxylic end-capped PNIPAAm (PNIPAAm–COOH) to AHA through amide bond linkages. AHA was prepared by grafting ADH to the HA backbone, and PNIPAAm–COOH copolymer was synthesized via a facile thermoradical polymerization technique by polymerization of NIPAAm using 4,40-azobis(4-cyanovaleric acid) as an initiator (Tan et al., 2009). This newly described thermoresponsive AHA-g-PNIPAAm copolymer demonstrated attractive properties to serve as cell or pharmaceutical delivery vehicles for a variety of TE applications.

A thiolated HA derivative, 3,30-dithiobis-(propanoic dihydrazide)-modified HA (HA-DTPH), was synthesized to fabricate nanofibrous scaffolds. Poly(ethylene oxide) (PEO) was blended with HA-DTPH as a viscosity modifier to facilitate the fiber formation during electrospinning.

A uniform HA-DTPH/PEO nanofibrous scaffold without beads was fabricated, which was cross-linked through poly(ethylene glycol) diacrylate (PEGDA)-mediated conjugate addition. PEO was subsequently extracted using DI water and an electrospun HA-DTPH nanofibrous scaffold was finally obtained. NIH3T3 fibroblasts attached to the scaffold and spread, demonstrating an extended dendritic morphology within the scaffold, which suggests potential applications of HA-DTPH nanofibrous scaffolds in cell encapsulation and tissue regeneration (Ji, Ghosh, Shu, et al., 2006). The effect of the electrospinning solvent on electrospinnability was investigated by Liu et al. (2011). The addition of formic acid greatly improved the electrospinnability of HA solution, and pure HA nanofibers with a mean diameter below 100 nm were fabricated under the optimal condition.

Two polysaccharides, LMW hyaluronic acid-1 (LMWHA-1) and LMW hyaluronic acid-2 (LMWHA-2), with their molecular weight of 1.45–105 and 4.52–104 Da, respectively, were prepared from HMW hyaluronic acid (HMWHA, 1.05–106 Da) by Ke et al. (2011). LMWHA-1, LMWHA-2, and HA were studied for their antioxidant activities. In *in vitro* antioxidant assay, LMWHA showed strong inhibition of lipid peroxidation and scavenging activities of hydroxyl radical, moderate 1,1-diphenyl-2-picrylhydrazyl radical, and superoxide anion scavenging activity. In addition, the LMWHA-1 exhibited much stronger antioxidant activity than LMWHA-2 and HA. They also proved that the administration of LMWHA was able to overcome cyclophosphamide (CY)-induced immunosuppression and significantly raised the activity of superoxide dismutase, catalase, and glutathione peroxidase and total antioxidant capacity in immunosuppressed mice.

The structural changes of gamma irradiated HA were studied by gel-permeation chromatography, viscosity, pH, Hunter color measurement, UV spectrophotometry, and FT-IR spectroscopy by Kim, Srinivasan, et al. (2008). The results demonstrated that GM decreased molecular weight size, viscosity, and pH of the hyaluronic acid and its color turned to intense yellow. UV spectra of the irradiated HA showed a change at 265 nm, which indicates the formation of double bonds. Differences in the height and shape of certain absorption bonds of FT-IR spectra in the range 1700–1750 cm^{-1} were also observed, which is associated with the formation of carboxylic acid. From these structural changes of the HA, GM may have a role in the formation of pyrancarboxylic acid rings. DPPH radical scavenging ability and the reducing power of gamma irradiated HA were significantly higher than that of nonirradiated HA.

5. APPLICATIONS OF HYALURONIC ACID

The unique viscoelasticity and limited immunogenicity of hyaluronic acid have led to its use in several biomedical applications, such as viscosupplementation in osteoarthritis (OA) treatment, as an aid in eye surgery, and for wound regeneration. In addition, HA has recently been explored as a drug delivery agent for different routes such as nasal, oral, pulmonary, ophthalmic, topical, and parenteral and also in TE applications.

5.1. Biomedical applications

Bacterial contamination of materials is of crucial importance in diverse fields such medical, food, or cosmetic industries. Once adhered on a surface, bacteria form colonies and subsequently biofilms that serve as reservoirs for the development of pathogenic infections. Bacterial biofilm infections are particularly problematic because sessile bacteria can withstand host immune responses and are drastically more resistant to antibiotics, biocides, and hydrodynamic shear forces than their planktonic counterparts (Mah & O'Toole, 2001). Prevention of biofilm formation is clearly preferable to any treatment strategy (Glinel, Thebault, Humblot, Pradier, & Jouenne, 2012). An efficient approach to prevent biofilm formation consists in immobilizing a bactericidal molecule on the support. Thus, various synthetic approaches based on the coating, grafting, or release of bactericidal substances such as metal derivatives, poly(ammonium salts), and antibiotics have been extensively explored to produce antimicrobial material.

Nisin (an antimicrobial peptide) has been attached to hyaluronic acid (HA) to obtain an antimicrobial biopolymer under solution or gel form. Various amounts of peptide have been grafted onto HA through a controlled reaction to obtain a covalently grafting by the formation of amide bonds. This modified HA exhibited a great antimicrobial property on the three tested bacterial species and proved as a potential material to avoid bacterial contamination in various applications as wound dressings, contacts lenses, cleaning solutions for contact lenses, and cosmetics formulations (Lequeux, Ducasse, Jouenne, & Thebault, 2014).

Diabetic foot ulcer (DFU) is one of the major complications associated with diabetic mellitus. Neuropathy and ischemia are two major etiologic factors leading to DFU (William & Keith, 2003). Intensive care should be taken for patients with DFU for the prevention of amputation. These ulcers tend to heal slowly, and healing can be complicated by polymicrobial infection and heavy exudate formation, which place these patients at a higher risk for limb amputation (Schwartz et al., 2013). Wound dressings with multiple performance parameters are needed for the treatment of DFU. A modern wound dressing should include many factors more than just exudate absorption. It should be able to keep up a moist interface between the wound and dressing, prevent infection, and create an optimal environment that supports healing with esthetically satisfactory scar (Hilton, Williams, Beuker, Miller, & Harding, 2004). An antimicrobial sponge composed of chitosan, hyaluronic acid (HA), and nanosilver (nAg) as a wound dressing for DFU infected with drug-resistant bacteria was prepared by Anisha, Biswas, Chennazhi, and Jayakumar (2013). nAg (5–20 nm) was prepared and mixed with chitosan and hyaluronic acid and characterized. The antimicrobial studies with selected bacteria suggest that these nanocomposite sponges could be used as a potential material for wound dressing for DFU infected with antibiotic-resistant bacteria such as *Escherichia coli*, *Staphylococcus aureus*, methicillin-resistant *S. aureus*, *Pseudomonas aeruginosa*, and *Klebsiella pneumonia* if the optimal concentration of nAg exhibiting antibacterial action with least toxicity toward mammalian cells was identified.

HA is an attractive starting material for the construction of bulk gels or hydrogel particles (Xu, Jha, Harrington, Farach-Carson, & Jia, 2012), but its applications in bone TE are limited by its poor mechanical properties. For this reason, it is often associated with calcium phosphates in order to obtain reinforced and/or injectable bone cements, consisting in HA gels containing hydroxyapatite (Nageeb et al., 2012) or calcium phosphate cements (CPCs) containing HA (Ahmadzadeh-Asl et al., 2011). Incorporating drug-loaded microspheres in mineral bone cements is an alternative strategy to improve their ability as drug delivery materials. To synthesize microspheres according to a reproducible process and control at the same time their morphology and their encapsulation efficiency is one of the main challenges of the conception of such drug-loaded bone substitute. In this context, we investigated the potentialities of two HAs, differing by their molecular weight, to form microspheres by a spray-drying technique. Erythrosin B was encapsulated as a model drug, and spray-drying process conditions were optimized by Fatnassi et al. (2013). HA molecular weight and concentration appeared to have a significant influence on process parameters and resulting microspheres. However, the introduction of HA microspheres in a mineral cement led to a sustained release, due to HA gel formation within the pores of the mineral matrix that decreases the diffusion rate. Hyaluronic acid microspheres could be of particular interest to formulate bone cements with extended ability to release active compounds.

A number of functional HA derivatives have also been developed, in order to modulate its biological properties (e.g., enzymatic degradability through esterification as in HYAFF-11 derivatives) and/or to prepare biomimetic three dimensional (3D)-extended matrices (e.g., hydrogels) or dispersible materials (e.g., nanoparticles). We are specifically interested in photopolymerization. This is a widespread method for the *in situ* preparation of TE matrices, whose biocompatibility and efficacy have been demonstrated in a number of studies (Bryant & Anseth, 2002).

Photopolymerization has been used for the preparation of HA/poly(ethylene glycol) (PEG) systems, whose degradability and mechanical properties can be controlled in a relatively independent fashion, respectively, through the HA and PEG content and molecular weight. We are specifically interested in photopolymerization. This is a widespread method for the in situ preparation of TE matrices, whose biocompatibility and efficacy have been demonstrated in a number of studies (Elisseeff et al., 2000). Photopolymerization has been used for the preparation of HA/PEG systems by Ouasti et al. (2011), whose degradability and mechanical properties were controlled in a relatively independent fashion, respectively, through the HA and PEG content and molecular weight.

Polymer conjugation has become an important strategy, but one challenge has been in controlling pharmacokinetics while maintaining affinity for the target (Gilli, Ferretti, & Gilli, 1994). For therapeutic proteins, conjugation of PEG is an established strategy for increasing circulation time, inhibiting enzymatic degradation, and improving solubility. PEG is an uncharged polyether with established solubility and biocompatibility (DeNardo et al., 2003). The benefits of PEGylation are that the PEG chain helps to protect the compound from enzymatic degradation and that the increased size decreases clearance rates. Additionally, the hydrophilic nature of the PEG chain can help to reduce aggregation and improve the solubility of the therapeutic. A study on a model peptide inhibitor of tumor necrosis factor-α to investigate the effects of site-specific conjugation to HA and PEG was carried out by Elder, Hannes, Atoyebi, and Washburn (2013). The results suggest that conjugation strategies involving both PEG and charged polymers, such as HA, could result in significant enhancements in the activities of therapeutic proteins.

Atherosclerosis is a chronic inflammatory condition of the blood vessel wall that can lead to arterial narrowing and subsequent vascular compromise. Although there are a variety of open and endovascular procedures used to alleviate the obstructions caused by atherosclerotic plaque, blood vessel instrumentation itself can lead to renarrowing of the vessel lumen through intimal hyperplasia, wound contracture, or a combination of the two. However, biologically active elements of the ECM are also important in the vascular remodeling that takes place in both atherosclerosis and renarrowing of the vessel lumen after instrumentation (Toole, Wight, & Tammi, 2001). One such element of the ECM that is important in both of these processes is hyaluronic acid (HA), otherwise known as hyaluronan (Bot, Hoefer, Piek, & Pasterkamp, 2008). HA is upregulated in areas of vascular injury and has been shown to increase VSMC migration and proliferation, key events in the progression of atherosclerosis and vascular renarrowing after surgical intervention (Benjamin Sadowitz, Keri Seymour, Vivian Gahtan, & Maier, 2012).

OA is a degenerative and debilitating disorder of diarthrodial joints affecting approximately 70% of 70-year-olds. It is associated with progressive damage of articular cartilage, leading to considerable pain resulting in loss of mobility and the requirement of continuous healthcare for patients (Yelin, 1992). The natural joint is lubricated by synovial fluid, which is highly viscous, enabling it to cushion and lubricate the joint. Hyaluronic acid (HA) is a major component of SF and articular cartilage and plays an important role in the lubrication of the cartilage surface in addition to helping maintain the structural resistance of cartilage to compressive forces. The application of HA and dipalmitoyl phosphatidylcholine onto damaged human cartilage resulted in improved lubrication between the cartilage surfaces on friction within a human cartilage damage model (Forsey et al., 2006).

CY is an anticancer and immunosuppressant drug that induces the production of reactive oxygen species (ROS). The excessive production of ROS plays multiple important roles in tissue damage and loss of function in a number of immune tissues and organs. For example, ROS can induce cell death by injuring the DNA of normal cells, resulting in the damage of the immune system (Diaz-Montero et al., 2012). It has been reported that polysaccharides can attenuate this oxidative damage of a tissue indirectly by enhancing natural defenses of cell and/or directly by raising the immunostimulatory activity (Chen et al., 2012; Wang et al., 2011). The immunostimulatory activities of two LMW hyaluronic acids (LMWHA-1 and LMWHA-2 with MW of 1.45–105 and 4.52–104 Da, respectively) and HA (MW, 1.05–106 Da) were evaluated using *in vitro* cell models and *in vivo* animal models, and their effects on angiogenesis were measured *in vivo* using the chick embryo chorioallantoic membrane assay by Ke et al. (2013). The results demonstrated that LMWHA-1, LMWHA-2, and HA could promote the splenocyte proliferation, increase the activity of acid phosphatase in peritoneal macrophages, and strengthen peritoneal macrophages to devour neutral red *in vitro* in a dose-dependent manner. Furthermore, LMWHA-1 and LMWHA-2 exhibited much stronger immunostimulatory activity than HA.

Technique commonly used in orthopedics is the insertion of prostheses in the body for the fixation or reconstruction of bones or their parts. Prostheses are generally made of biocompatible metals (in particular titanium and cobalt chrome), polymers, ceramics, hydroxyapatite, or their combinations (e.g., metals coated with a layer of hydroxyapatite). Bacterial infections due to implanted prosthesis still represent a serious complication in orthopedic surgery. Studies indicate that the procedures for implanting a prosthesis and the presence of the prosthesis itself in the site of bone fracture damage the response of the local immune system with the result that the number of bacteria required to cause an infection can fall by a factor of even 10,000 (Flückiger & Zimmerli, 2000). Several researchers have proposed antibacterial materials with nonfouling properties, in particular for use as coatings of the orthopedic prostheses; such materials should preferably be capable to release the drug immediately after the surgical operation and at least during the following 6 h, preferably up to 48–72 h, so as to cover the critical period of possible bacterial attack and proliferation in the intervention site (Yeap et al., 2006). Physical hydrogels have been obtained by Giammona et al. (2013) from hyaluronic acid derivatized with polylactic acid in the presence or in the absence of PEG chains. They have been extemporarily loaded with antibacterial agents, such as vancomycin and tobramycin. These medicated hydrogels have been used to coat titanium disks (chosen as simple model of orthopedic prosthesis), and *in vitro* studies in simulated physiological fluid have been performed as a function of time and for different drug loading and polymer concentration values. Obtained results suggest the potential use of these hydrogels in the orthopedic field, in particular for the production of antibacterial coatings of prostheses for implant in the human or animal body in the prevention and/or treatment of postsurgical infections.

Organic–inorganic materials, combining functional properties of inorganic compounds and polymers, are of significant interest for biomedical applications. Composite films, containing halloysite nanotubes (HNTs), hydroxyapatite (HAp), and biocompatible polymers, are currently under intensive investigation. Electrophoretic deposition (EPD) is an attractive method for the deposition of composite films for biomedical applications.

This method is widely used for the deposition of inorganic materials, polymers, and composites. EPD is based on the electrophoretic motion of colloidal particles or polymer macromolecules under the influence of an electric field and deposit formation at the electrode surface (Zhitomirsky, 2002). Anodic EPD of composite films, containing HNT and HA in an HYH matrix, was carried out by Deen and Zhitomirsky (2014). The results demonstrated that HYH can be used as efficient charging and dispersing agent for HNT and HA in suspensions and film-forming agent for EPD of HNT–HA–HYH films.

The use of osteotransductive CPCs in vertebroplasty is limited by their low injectability and disintegration. Liquid-phase separation, termed filter pressing, is a frequent problem for cement injectability (Bohner, Gbureck, & Barralet, 2005). Viscosity-enhancing agents such as hyaluronic acid and chondroitin-4-sulfate are used to improve the injectability of brushite (chronOS Inject, Synthes) and apatite (Biopex, Mitsubishi Materials Corporation) cements, respectively. The adhesiveness of brushite cements due to the presence of sodium hyaluronate supported intimate contact between the cement and bone surface (Apelt et al., 2004). Moreover, the addition of hyaluronic acid to an injectable brushite cement had little effect on its osteoconductive properties, except for a light decrease in initial resorption rate (Flautre, Lemaitre, Maynou, Van Landuyt, & Hardouin, 2003).

Recently, attention has been paid to the use of microneedles fabricated from biocompatible and biodegradable polymers (Donnelly et al., 2011; Jin, Han, Lee, & Choi, 2009) and carbohydrates (Ito, Murano, Hamasaki, Fukushima, & Takada, 2011), which are free from the risk of complications. If left in the skin, these types of needles safely degrade and eventually disappear. They also have the potential for loading drugs into a matrix of needles and releasing them in the skin by biodegradation or dissolution in the interstitial fluid: a one-step application. Hyaluronic acid (HA) was used (Liu et al., 2014) to fabricate novel dissolving microneedle arrays in this study. HA is a water-soluble polymer of disaccharides, naturally found in many tissues, such as the skin, the cartilage, and the vitreous humor. In 2003, the FDA approved HA injections for filling soft tissue defects. This study indicated that self-dissolving HA microneedle arrays are a useful alternative to improve the transdermal delivery of drugs, especially drugs with relatively HMW without seriously damaging the skin. The HA-fabricated microneedle arrays containing alendronate and insulin were found be effective for improving the transdermal drug delivery.

5.2. TE applications

TE is a field of research that is aimed at regenerating tissues and organs (Daamen et al., 2003). Cells, scaffolds, and growth factors are the three main components for creating a tissue-engineered construct. The principal aim of a scaffold design should be to mimic the native ECM of the target tissue as much as possible (Ma, Gao, Gong, & Shen, 2005). The development of biodegradable polymers to perform the role of a temporary matrix is an important factor in the success of cell transplantation. Cells, scaffold and growth stimulating signals are generally refer to as the tissue engineering triad, the key components of engineered tissues.

Collagen type I, a major protein of the ECM in mammals, is a suitable scaffold material for regeneration. Another important constituent of the ECM, hyaluronic acid (hyaluronan, HA), has been used for medical purposes due to its hydrogel properties and biodegradability. Chitosan is a linear polysaccharide composed of b1–b4-linked D-glucosamine residues, and its potential as a biomaterial is based on its cationic nature and high charge density in solution. A study was conducted to evaluate the characteristics of scaffolds composed of different ratios of type I comb collagen and chitosan with added HA in order to obtain the optimum conditions for the manufacture of collagen–hyaluronan–chitosan (Col–HA–Ch; comprising collagen (Lin et al., 2009). Overall, the 9:1:1 mixing ratio of collagen, hyaluronan, and chitosan was observed to be optimal for the manufacture of complex scaffolds. Furthermore, Col–HA–Ch tripolymer scaffolds, especially Col9HACh1, could be developed as a suitable scaffold material for TE applications.

5.2.1 Lung TE applications

Diseases like pulmonary hypoplasia (found in neonates) and emphysema (a chronic lung disease) have a deficient alveolar epithelium, tissue loss, or reduced alveolar surfactant synthesis. In all these instances, TE represents an attractive potential for regeneration or augmentation with engineered functional pulmonary tissue. Several strategies are being adopted to evolve suitable scaffolds for lung TE. There is a growing interest in blending natural and synthetic polymers as biomaterials for creating complex structures, which will act as scaffolds for TE. Turner, Kielty, Walker, and Canfield (2004) and Amarnath, Srinivas, and Ramamurthi (2006) had shown a commercial benzyl ester of HA and laboratory cross-linked hylan as excellent biomaterials for the promotion of adherence of vascular endothelial cells and vascular TE.

In spite of many potential applications of HA, it has some inherent drawbacks. Chemical modification through grafting has received considerable attention in the area of biomedical applications. Poly(HEMA) is one of the most important hydrogels in the biomaterials world since it has many advantages over other hydrogels (Pescosolido et al., 2011). These include a water content similar to living tissue, inertness to biological processes, resistance to degradation, permeability to metabolites, and resistance to absorption by the body. Hence, a graft copolymer of HA and poly(HEMA) appeared as a good choice for the synthesis of a natural–synthetic polymer hybrid matrix for use as a scaffold for lung TE.

An "*in situ*" biodegradable gel consisting of chitosan, glycerol phosphate (GP), and oxidized hyaluronic acid (HDA) was synthesized and characterized by Nair, Remya, Remya, and Nair (2011). This is a two-component hydrogel system where chitosan neutralized with GP resulted in instantaneous gelling when combined with HDA. The gels are cytocompatible and could be freeze-dried to form porous scaffolds. The percentage porosity of the freeze-dried chitosan hyaluronic acid dialdehyde gels (CHDA) increased with increasing oxidation. Fibroblast cells seeded onto CHDA porous scaffolds adhered, proliferated, and produced ECM components on the scaffold. Chondrocytes encapsulated in CHDA gels retained their viability and specific phenotypic characteristics. The gel material can be a good scaffold and encapsulating material for TE applications.

5.2.2 Bone TE applications

Titanium (Ti) and its alloy are widely used in the biomedical field, such as hip joint replacement devices and heart valves, due to their excellent corrosion resistance and mechanical properties (Geetha, Singh, Asokamani, & Gogia, 2009). Among the titanium alloys employed, Ti–Nb–Zr alloy is a promising candidate material because of its low elastic modulus and shape memory effect (Li et al., 2011). The bare inert surface of titanium (Ti) alloy typically causes early failures in implants. LBL self-assembly is one of the simple methods for fabricating bioactive multilayer coatings on titanium implants. Zhang, Li, Yuan, Cuia, and Yang (2013) prepared a dopamine-modified hyaluronic acid/chitosan (DHA/CHI) bioactive multilayer built on the surface of Ti–24Nb–2Zr (TNZ) alloy. Zeta potential oscillated between −2 and 17 mV for DHA- and CHI-ending layers during the assembly process, respectively. Preosteoblast MC3T3-E1 cells were cultured on the original TNZ alloy and TNZ/(DHA/CHI)5 to evaluate the effects of DHA/CHI multilayer on osteoblast proliferation *in vitro*. The proliferation of osteoblasts on TNZ/(DHA/CHI)5 was significantly higher than that on the original TNZ alloy. The results of this study indicate that the proposed technique improves the biocompatibility of TNZ alloy and can serve as a potential modification method in orthopedic applications.

Recent studies suggest that bone marrow stromal cells are a potential source of osteoblasts and chondrocytes and can be used to regenerate damaged tissues using a TE approach. However, these strategies require the use of an appropriate scaffold architecture that can support the formation *de novo* of either bone and cartilage tissue, or both, as in the case of osteochondral defects. A novel hydroxyapatite/chitosan (HA/CS) bilayered scaffold was developed by Oliveira et al. (2006) by combining a sintering and a freeze-drying technique and aims to show the potential of such type of scaffolds for being used in TE of osteochondral defects. Results have shown that materials do not exert any cytotoxic effect. Complementarily, *in vitro* (phase I) cell culture studies were carried out to evaluate the capacity of HA and CS layers to separately support the growth and differentiation of goat marrow stromal cells into osteoblasts and chondrocytes, respectively. The obtained results concerning the physicochemical and biological properties of the developed HA/CS bilayered scaffolds show that these constructs exhibit great potential for their use in TE strategies, leading to the formation of adequate tissue substitutes for the regeneration of osteochondral defects.

Hyaluronic acid (HA) functionalized with ethylenediamine (EDA) has been employed to graft α-elastin in different proportions by Palumbo et al. (2013). In particular, an HA-EDA derivative bearing 50 mol% of pendant amino groups has been successfully employed to produce the copolymer HA-EDA-g-α-elastin containing 32% w/w of protein. After grafting with α-elastin, the remaining free amino groups reacted with ethylene glycol diglycidyl ether (EGDGE) for producing chemical hydrogels, proposed as scaffolds for TE. The presence of α-elastin grafted to HA-EDA improves attachment, viability, and proliferation of primary rat dermal fibroblasts and human umbilical artery smooth muscle cells. Biological performance of HA-EDA-g-α-elastin/EGDGE scaffold is comparable to that of a commercial collagen type I sponge (Antema®), chosen as a positive control.

The feasibility of hyaluronic acid/sodium alginate (HA/SA) scaffold-based interpenetrating polymeric network (IPN) for the proliferation and chondrogenic differentiation of the human adipose-derived stem cells (hADSCs) was evaluated by Son et al. (2013). The hADSCs cultured in HA/SA IPN scaffold exhibited enhanced cell adhesion and proliferation compared with the HA scaffold. Superior chondrogenic differentiation of hADSCs in HA/SA IPN scaffold, compared with HA-based scaffold, was confirmed by measuring expression levels of chondrogenic markers.

5.2.3 Stem cells for TE applications

Stem cells have the ability to self-renew and differentiate into a wide range of specialized cell types. Thus, they are very promising for the regeneration of aged, injured, and diseased tissues (Kim & De Vellis, 2009). Embryonic stem cells (ESCs), induced pluripotent stem cells, and adult stem cells are currently the primary cell source for research in the lab and clinic. ESCs, which are derived from the inner cell mass of early-stage embryos, can be differentiated into most of the cell types found in the body and can be expanded *in vitro* (Lumelsky et al., 2001). Hyaluronic acid (HA) hydrogels were synthesized that could be degraded through a combination of cell-released enzymes and used them to culture mouse mesenchymal stem cells. To form the hydrogels, HA was modified to contain acrylate groups and cross-linked through Michael addition chemistry using nondegradable, plasmin degradable, or matrix metalloproteinase degradable cross-linkers. Cells in stiffer hydrogels showed less spreading, migration, and slower proliferation rates. The information gained can provide valuable insight into the designing of hydrogels for 3D stem culture.

Mesenchymal stem cells (MSCs) are commonly used in TE applications due to their availability, ability to expand, and capacity to differentiate into multiple cell types. An important clinical application for MSCs under widespread investigation is cartilage repair. Mature hyaline cartilage is vascular and lymphatic, with cells comprising only about 5% of the tissue volume (Chung & Burdick, 2008). Current clinical methods to repair defective cartilage are limited in their ability to regenerate functional cartilage in terms of both composition and mechanics (Ahmed & Hincke, 2010). Due to these shortcomings, recent research has focused on the use of TE approaches to repair cartilage tissue.

5.2.4 Cartilage TE applications

The development of hydrogels tailored for cartilage TE has been a research and clinical goal for over a decade. Directing cells toward a chondrogenic phenotype and promoting new matrix formation are significant challenges that must be overcome for the successful application of hydrogels in cartilage tissue therapies. Gelatin–methacrylamide hydrogels have shown promise for the repair of some tissues but have not been extensively investigated for cartilage TE. Levett et al. (2014) encapsulated human chondrocytes in

Gel–MA-based hydrogels and show that with the incorporation of small quantities of photocross-linkable hyaluronic acid methacrylate, and to a lesser extent chondroitin sulfate methacrylate, chondrogenesis and mechanical properties can be enhanced.

The response of MSCs to a matrix largely depends on the composition as well as the extrinsic mechanical and morphological properties of the substrate to which they adhere to. Collagen–GAG scaffolds have been extensively used in a range of TE applications with great success. This is due in part to the presence of the GAGs in complementing the biofunctionality of collagen. In this context, the overall goal of this study was to investigate the effect of two GAG types: chondroitin sulfate and hyaluronic acid (HA) on the mechanical and morphological characteristics of collagen-based scaffolds and subsequently on the differentiation of rat MSCs *in vitro*. Morphological characterization revealed that the incorporation of HyA resulted in a significant reduction in scaffold mean pore size (93.9 µm) relative to collagen–CS (CCS) scaffolds (136.2 µm). In addition, the collagen–HyA (CHyA) scaffolds exhibited greater levels of MSC infiltration in comparison with the CCS scaffolds. These CHyA scaffolds show great potential as appropriate matrices for promoting cartilage tissue repair.

A procedure to obtain electrospun mats of hyaluronic acid (HA) stable in aqueous media in one single step has been developed. It consists in combining an HA solution with a divinyl sulfone one as cross-linker in a three-way valve to immediately electroblow their mixture. Therefore, it is necessary to cross-link it to obtain a material stable in aqueous conditions. Many attempts have been done in this line, consisting in either a physical (Wang et al., 2005) or a chemical (Xu et al., 2009) cross-linking after the electrodeposition process, in order to obtain HA mats insoluble in water. In all these works, the cross-linking was performed in a second step following the electrospinning process of the HA mat. In another approach, a dual-syringe setup was employed to combine thiolated derivatives of HA mixed with PEO with PEGDA as cross-linker agent (Ji, Ghosh, Li, et al., 2006) during the electrospinning process, but an additional second step was still required to remove the PEO.

Still, important disadvantages of these membranes are their reduced thickness and their high equilibrium water content, which imply lack of mechanical properties and limited manageability. These drawbacks could become a problem for some TE applications. This led us to combine the HA mat with another polymer into a two-layer membrane. Here, HA has been electrospun onto previously obtained mats of poly(L-lactic acid)

(PLLA), which possesses a great electrospinning processability, proper mechanical behavior, and manipulability and is a noncytotoxic and biodegradable FDA-approved polymer (Cheung, Lau, Lu, & Hui, 2007). Arnal-Pastor, Martínez Ramos, Pérez Garnés, Monleón Pradas, and Vallés Lluch (2013) developed a coelectroblowing procedure that allows the fabrication of insoluble HA nanofiber mats in a single step. It consists in mixing an HA solution with another of the cross-linker in a three-way valve just before electrospinning. These nanofibers can be electroblowed stably on previously electrospun tougher PLLA mats to obtain bilayered membranes. The flexibility of the fabrication process allows the preparation of membranes with different thicknesses and properties by varying the electrospinning times.

5.2.5 Heart TE applications

There are about 1.4 million arterial bypass operations performed using vascular grafts in the United States every year. Although the large size (6 mm) vascular grafts have been satisfactory in clinic applications, suitable small caliber (6 mm) vascular grafts are still insufficient to meet the existing clinical needs due to improper endothelialization, poor biocompatibility, and low mechanical properties. Therefore, there is a need to address the structural design and chemical composition of vascular grafts to develop an optimal small size vascular graft (Seidlits et al., 2011). An intima layer scaffold of the blood vessel for endothelialization was prepared by Zhu, Fana, and Wang (2014) using novel human-like collagen/hyaluronic acid (HLC/HA) composite at different mass ratios of 40/1, 20/1, and 10/1 by freeze-drying process. The structure, mechanical strength, degradation, and biocompatibility of the vascular HLC/HA scaffold were evaluated. The results showed that the 10/1 HLC/HA composited an optimal scaffold with (1) an interconnected porous network with a pore diameter of 12 ± 2 µm and porosity of 89.3%; (2) better mechanical properties with higher stress of 321.7 ± 15 kPa and strain of $45.5\pm0.2\%$ than 40/1, 20/1, and pure HLC scaffolds; (3) only 9% degradation upon immersion in PBS for 45 days at 37 °C *in vitro*; and (4) excellent biocompatibility. This study suggests that the 10/1 HLC/HA composite has a broad prospect of application as luminal vascular scaffold in the TE.

Dahlmann et al. (2013) developed a fully defined *in situ* hydrogelation system based on alginate (Alg) and hyaluronic acid (HA), in which their aldehyde and hydrazide derivatives enable covalent hydrazone cross-linking of polysaccharides in the presence of viable myocytes. A combination of HyA and highly purified human collagen I led to significantly increased active contraction force compared with collagen only. Therefore, our *in situ* cross-linking hydrogels represent a valuable toolbox for the fine-tuning of engineered cardiac tissue's mechanical properties and improved functionality, facilitating clinical translation toward therapeutic heart muscle reconstruction.

Fibrin gel is widely used as a TE scaffold. However, it has poor mechanical properties, which often result in rapid contraction and degradation of the scaffold. An IPN hydrogel composed of fibrin and hyaluronic acid–tyramine (HA–Tyr) was developed by Lee and Kurisawa (2013) to improve the mechanical properties. The fibrin network was formed by cleaving fibrinogen with thrombin, producing fibrin monomers that rapidly polymerize. The HA network was formed through the coupling of tyramine moieties using horseradish peroxidase and hydrogen peroxide (H_2O_2). The degree of cross-linking of the HA–Tyr network can be tuned by varying the H_2O_2 concentration, producing IPN hydrogels with different storage moduli. Cell proliferation and capillary formation occurred in IPN hydrogels, which suggests that fibrin–HA–Tyr IPN hydrogels are a potential alternative to fibrin gels as scaffolds for TE applications that require shape stability.

5.2.6 Brain TE applications

A hyaluronic acid–collagen (HA–Col) sponge with an open porous structure and mechanical behavior comparable to brain tissue was developed by Wang and Spector (2009). HA–Col scaffolds with different mixing ratios were prepared by a freeze-drying technique and cross-linked with water-soluble carbodiimide to improve mechanical stability. Certain features of the mechanical properties of HA–Col scaffolds prepared with a Col/HA mixing ratio of 1:2, and pure HA sponges, were comparable to brain tissue.

Bioactive, *in situ*-forming materials have the potential to complement minimally invasive surgical procedures and enhance tissue healing. For such biomaterials to be adopted in the clinic, they must be cost-effective and easily handled by the surgeon and have a history of biocompatibility. A novel and facile self-assembling strategy to create membranes and encapsulating structures using collagen and hyaluronic acid (HA) was created by Chung, Jakus, and Shah (2013). Unlike membranes built by LBL deposition of oppositely charged biomolecules, the collagen–HA membranes described here form a diffusion barrier upon electrostatic interaction of the oppositely charged biomolecules, which is further driven by osmotic pressure imbalances. The resulting membranes have a nanofibrous architecture, thicknesses of 130 lm, and a tensile modulus (0.59 ± 0.06 MPa) that can increase sevenfold using carbodiimide chemistry (4.42 ± 1.46 MPa). Collagen–HA membranes support MSC proliferation and have a slow and steady protein release profile (7% at day 28), offering opportunities for targeted tissue regeneration.

5.2.7 Dermal TE applications

The fabrication of new dermal substitutes providing mechanical support and cellular cues is urgently needed in dermal reconstruction. Yan et al. (2013) prepared silk fibroin/chondroitin sulfate/hyaluronic acid (HA) ternary scaffolds (95–248 μm in pore diameter and 88–93% in porosity by freeze-drying). By the incorporation of CS and HA with the SF solution, the chemical potential and quantity of free water around ice crystals could be controlled to form smaller pores in the SF/CS/HA ternary scaffold main pores and improve scaffold equilibrium swelling. This feature offers benefits for cell adhesion, survival, and proliferation. *In vivo* SF, SF/HA, and SF/CS/HA (80/5/15) scaffolds as dermal equivalents were implanted onto dorsal full-thickness wounds of Sprague-Dawley rats to evaluate wound healing. Compared with SF and SF/HA scaffolds, the SF/CS/HA (80/5/15) scaffolds promoted dermis regeneration, related to improved angiogenesis and collagen deposition.

Macroporous elastic scaffolds containing gelatin (4% or 10%) and 0.25% hyaluronic acid (HA) were fabricated by Chang, Liao, and Chen (2013) by cryogelation for application in adipose TE. These cryogels have interconnected pores (200 lm), a high porosity (>90%), and a high degree of cross-linking (>99%). The higher gelatin concentration reduced the pore size, porosity, and swelling ratio of the cryogel but improved its swelling kinetics. Compressive mechanical testing of cryogel samples demonstrated nonlinear stress–strain behavior and hysteresis loops during loading–unloading cycles but total recovery from large strains. The presence of more gelatin increased the elastic modulus, toughness, and storage modulus and yielded a cryogel that was highly elastic, with a loss tangent equal to 0.03. Porcine adipose-derived stem cells (ADSCs) were seeded in the cryogel scaffolds to assess their proliferation and differentiation. *In vitro* studies demonstrated a good proliferation rate and the adipogenic differentiation of the ADSCs in the cryogel scaffolds, as shown by their morphological change from a fibroblast-like shape to a spherical shape, decreased actin cytoskeleton content, growth arrest, secretion of the adipogenesis marker protein leptin, Oil Red O staining for triglycerides, and expression of early (LPL and PPARc) and late (aP2 and leptin) adipogenic marker genes. *In vivo* studies of ADSCs/cryogel constructs implanted in nude mice and pigs demonstrated adipose tissue and new capillary formation; the expression of PPARc, leptin, and CD31 in immunostained explants; and the continued expression of adipocyte-specific genes. Both the *in vitro* and *in vivo* studies indicated that the gelatin/HA cryogel provided a structural and chemical environment that enabled cell attachment and proliferation and supported the biological functions and adipogenesis of the ADSCs.

5.3. Drug delivery applications

Drug release is another interesting application and formulations of HA, and its derivatives have been developed as topical, injectable, and implantable vehicles for the controlled and localized delivery of biologically active molecules (Vasiliu, Popa, & Rinaudo, 2005). HA has also been shown to have an antiplatelet activity (Burns & Valeri, 1996), which is important in avoiding thrombus formation; hence, HA is also used as a coating for blood-contacting implants.

A system that formulates or device that delivers therapeutic agent(s) to the desired body location(s) and/or provides timely release of therapeutic agent(s), such a system by which a drug is delivered, can have a significant effect on its efficacy. On the other hand, the very slow progress in the efficacy of the treatment of severe diseases has suggested a growing need for a multidisciplinary approach to the delivery of therapeutics to targets in tissues. From this, new ideas on controlling the pharmacokinetics, pharmacodynamics, nonspecific toxicity, immunogenicity, biorecognition, and efficacy of drugs were generated. These new strategies, often called drug delivery systems (DDS), are based on interdisciplinary approaches that combine pharmaceutics, polymer science, analytical chemistry, bioconjugate chemistry, and molecular biology (Allen, 2002).

Regulated drug release from biodegradable polymer matrices has been widely examined in order to release dispersed or dissolved drug in proportion to degradation of the polymer matrix. Recently, several types of biodegradable polymers have been extensively studied to obtain drug release that is responsive to a biological and external stimulus (Heller, 1985). A number of studies reported that hyaluronic acid (HA) is capable of being used as a drug delivery agent along various administration routes, including ophthalmic, nasal, pulmonary, parenteral, and topical applications (Yadav, Mishra, & Agrawal, 2008).

The benefits of using HA as a drug delivery vehicle are that it is biocompatible, nontoxic, noninflammatory, and biodegradable; it can efficiently function as a "homing device" because the HA receptor cluster determinant 44 (CD44) is overexpressed in many types of tumor cells; it provides protection to its "cargo"; and it imparts solubility to hydrophobic drugs.

Despite phenomenal advances in the inhalable, injectable, transdermal, nasal, and other routes of administration, the unavoidable truth is that oral drug delivery remains well ahead of the pack as the preferred delivery route (Masaoka, Tanaka, Kataoka, Sakuma, & Yamashita, 2006). Oral delivery continues to be the most popular route of administration due to its versatility, ease of administration, and probably most importantly patient compliance. HA is used as a carrier in various oral formulations such as microspheres and complexes to improve the solubility and bioavailability of poorly water-soluble drugs (Piao et al., 2007); however, several limitations still exist and influence the gastrointestinal absorption of exogenous HA when it is administered orally, such as its relatively HMW and poor liposolubility (Huang, Ling, & Zhang, 2007).

Nanochitosan (NC)-capturing hyaluronic acid (HA) multilayer films were reported by Park et al. (2012). The films were designed to deliver various biomaterials, such as charged, uncharged, hydrophobic, and hydrophilic molecules such as peptides and proteins. Biodegradable NC-capturing HA multilayer films are expected to perform dual roles within a single platform as capturing NCs at desired amount within the films as well as the multilayer film buildup capability. This drug-loaded multilayered system demonstrates a considerable promise for drug-eluting systems to prevent restenosis. The HA multilayer film capturing paclitaxel (PTX)-loaded NCs efficiently induced the hSMC apoptosis and inhibited the hSMC proliferation, showing potentials of the HA multilayer films as effective DDS.

Hyaluronan-based hydrogels formulated to include heparin (Heprasil™) with similar gels without heparin (Glycosil™) were compared for their ability to deliver bioactive BMP-2 *in vitro* and *in vivo*. The osteogenic activity of BMP-2 released from the hydrogels was evaluated by monitoring alkaline phosphatase activity and SMAD 1/5/8 phosphorylation in mesenchymal precursor cells. The osteoinductive ability of these hydrogels was determined in a rat ectopic bone model by 2D radiography, 3D m-CT, and histological analyses at 8 weeks postimplantation. Both hydrogels sustain the release of BMP-2. Importantly, the inclusion of a small amount of heparin (0.3%, w/w) attenuated release of BMP-2 and sustained its osteogenic activity for up to 28 days. In contrast, hydrogels lacking heparin released more BMP-2 initially but were unable to maintain BMP-2 activity at later time points. Ectopic bone-forming assays using transplanted hydrogels emphasized the therapeutic importance of the initial burst of BMP-2 rather than its long-term osteogenic activity. Thus, tuning the burst release phase of BMP-2 from hydrogels may be advantageous for optimal bone formation.

5.4. Gene delivery applications

Gene therapy has greatly advanced as a promising therapeutic tool for treating genetic disorders as well as tumors in the past decade. However, the bottleneck for gene therapy is delivery. The natural anionic polysaccharide hyaluronic acid (HA) was modified by introducing reduction-sensitive disulfide bond between the carboxyl groups and the backbone of HA (HA–SS–COOH). HA–SS–COOH and its corresponding unmodified stable analog HA were used to shield DNA/PEI (DP) polyplexes to form ternary complexes (DPS and DPH complexes) by He et al. (2013). In this study, reducible shielding (HA–SS–COOH) and stable hyaluronic acid shielding were introduced to the formation of DNA/PEI complexes via electrostatic interaction. The presence of HA–SS–COOH and HA coating showed lower cytotoxicity, higher gene transfection efficiency, and greatly enhanced cellular uptake by HA receptor overexpressed carcinoma cells. More importantly, HA–SS–COOH shielding was superior to HA due to the extra reduction-responsive deshielding function. Therefore, dual functional hyaluronic acid derivative-modified DNA/PEI ternary complexes may have an advantage in targeted gene delivery to cancer cells.

HA derivative carrying dialdehyde (HAALD)–α,β-polyaspartylhydrazide (PAHy) hydrogels were synthesized in phosphate-buffered saline (PBSA) solution by Zhang, Huang, Xue, Yang, and Tan (2011) and characterized through different methods including gel content and swelling, Fourier transformed infrared spectra, thermogravimetric analysis, *in vitro* degradation, and biocompatibility experiments. A scanning electron microscope viewed the interior morphology of HAALD–PAHy gel whose porous 3D structure enabled it to efficiently encapsulate the proteins. Sustained and stable protein release from the HAALD–PAHy hydrogel was observed during *in vitro* delivery experiments and exhibited its potentially high-application prospect in the field of protein drug delivery.

Although the protective and impermeable qualities of the skin protect the organism from losing water, minerals, and dissolved proteins, percutaneous delivery system has been taken advantage as a topical delivery system in pharmaceutical or cosmetic industries. In contrast with traditional drug administration pathway, transdermal administration is featured by its noninvasive procedure, which eliminates side effects and increases patient compliance and possibility for continuous and controlled drug absorption. Besides being a vital organ, the skin must be nourished as the other organs of the body by means of cosmetic formulations (Souto & Muller, 2008). HA nanoemulsion targeting to be applied as transdermal carrier for active lipophilic ingredient has been developed in previous studies, whose stability and delivery potential have been verified. The results suggested nanoemulsions could be successfully used as percutaneous delivery vehicle of active lipophilic ingredient, and HA has a favorable role in skin care for drug and cosmetic applications.

A variety of nanostructures composed of polyamino acid polyarginine (PArg) and polysaccharide hyaluronic acid (HA) as a preliminary stage were prepared by Oyarzun-Ampuero, Goycoolea, Torres, and Alonso (2011) before evaluating their potential application in drug delivery. PArg was combined with HMWHA or LMWHA to form nanoparticles by simply mixing polymeric aqueous solutions at room temperature. The results showed that the molecular weight of HA is a crucial determinant of formulation stability during mechanical isolation and in physiological conditions. This knowledge is useful not only for systems comprising PArg and hyaluronic acid but also for systems composed of other polymers. Further studies testing the potential of these systems as mucoadhesive nanocarriers for targeted drug delivery will be carried out, and the *in vitro–in vivo* behavior of these systems will be also evaluated.

5.5. Targeted drug delivery

The prolonged circulation of the nanoparticles in the bloodstream may increase their probability of reaching the tumor tissue after systemic administration *in vivo*, which is due to the abnormal characteristics of tumors, such as the fenestrated vasculature and the lack of a lymphatic drainage system. However, this passive tumor targeting strategy is partially limited in the effective diagnosis or therapy of tumors, because conventional nanoparticles do not have the ability to be specifically internalized into the target cells, resulting in the release of a significant portion of the payloads at the extracellular phase. To overcome this limitation, considerable efforts have been made to develop nanoparticles capable of actively targeting cancer cells. Such nanoparticles have been modified with targeting moieties, such as antibodies, proteins, and various ligands that can selectively bind to receptors overexpressed on the target cells. Hyaluronic acid (HA), a natural polysaccharide found in the ECM and synovial fluids of the body, has been investigated as a targeting moiety of the drug conjugates or nanoparticles for cancer therapy because it can specifically bind to various cancer cells that overexpress CD44, an HA receptor. The PEGylation of hyaluronic acid nanoparticles improves tumor targetability *in vivo* (Choi et al., 2011).

PEG-conjugated hyaluronic acid–ceramide (HACE) was synthesized by Cho et al. (2012) for the preparation of doxorubicin (DOX)-loaded HACE-PEG-based nanoparticles, 160 nm in mean diameter with a negative surface charge. Greater uptake of DOX from these HACE-PEG-based nanoparticles was observed in the CD44 receptor highly expressed SCC7 cell line, compared with results from the CD44-negative cell line, NIH3T3. A strong fluorescent signal was detected in the tumor region upon intravenous injection of cyanine 5.5-labeled nanoparticles into the SCC7 tumor xenograft mice; the extended circulation time of the HACE-PEG-based nanoparticle was also observed. Pharmacokinetic study in rats showed a 73.0% reduction of the *in vivo* clearance of DOX compared to the control group. The antitumor efficacy of the DOX-loaded HACE-PEG-based nanoparticles was also verified in a tumor xenograft mouse model. DOX was efficiently delivered to the tumor site by active targeting via HA and CD44 receptor interaction and by passive targeting due to its small mean diameter (<200 nm). Moreover, PEGylation resulted in prolonged nanoparticle circulation and reduced DOX clearance rate in an *in vivo* model.

Slightly modified HA derivatives were used for target-specific intracellular delivery of nucleotide therapeutics, and highly modified HA derivatives were used for long-acting conjugation of peptide and protein therapeutics. The chemical modifications were carried out through carboxyl groups of HA, because carboxyl groups of HA are known to be the recognition sites for HA receptors and hyaluronidase (Banerji et al., 2007). The chemical modification of HA–COOH would change its biological behaviors in the body. As an example, it was reported that enzymatic degradation of HA derivatives was delayed with increasing degree of HA modification. HA can be designed to have various functional groups in the pendant groups. In order to introduce amine groups to HA, ADH, hexamethylenediamine (HMDA), or cystamine can be grafted by the conjugation reaction with carboxyl groups of HA.

Target-specific intracellular delivery of siRNA is one of the most important issues for the development of siRNA therapeutics with remarkable potentials. Despite extensive research efforts, there are still lots of unsolved problems for effective delivery of siRNA. HA has been used as a targeting moiety of gene delivery carriers. HA receptors are abundant in some specific tissues, such as the liver, kidney, and most of cancer tissues. While a widely used tethering molecule of PEG cannot interact with cell membrane and is hard to go through the cell membrane, HA can bind to the receptor on the cell surface and be uptaken to the cells by HA receptor-mediated endocytosis. For example, HA–PLL conjugate was synthesized targeting HARE receptor of sinusoidal epithelial cells in the liver. They conjugated the reducing end of HA with ε-amino groups of PLL by reductive amination to synthesize the comb-type copolymer. This copolymer was used to make complexes with DNA, which were intravenously injected to animal models.

5.6. HA hydrogels

Hydrogels have recently drawn great attention for use in a wide variety of biomedical applications such as cell therapeutics, wound healing, cartilage/bone regeneration, and the sustained release of drugs. This is due to their biocompatibility and the similarity of their physical properties to natural tissue. Hydrogels are 3D networks of cross-linked hydrophilic polymers that typically show a high degree of swelling in aqueous environments without dissolution of polymeric networks. Among the various fabrication methods, photopolymerization is a very good means to prepare hydrogels due to its many advantages. Photopolymerization has recently received increased attention due to its capacity to allow for mixing of aqueous macromer solutions containing cells and bioactive factors. This mixture can be delivered in a minimally invasive manner and then rapidly cross-linked in physiological conditions *in situ* following brief exposure to UV light (Hutchison, Stark, Hawker, & Anseth, 2005). A photocured HA hydrogel containing an osteogenesis-inducing growth factor, GDF-5. HA hydrogels I–III were prepared by Bae et al. (2014) and confirmed to have controlled GDF-5 release profiles. Cytotoxicity and cell viability suggest that GDF-5-loaded HA hydrogel has proper biocompatibility for use as a scaffold, which can induce osteogenesis.

The protocols to prepare HA hydrogels can be classified into three types, direct cross-linking of HA, cross-linking of HA derivatives, and cross-linking between two different kinds of HA derivatives. According to the molar ratio of diamine to carboxyl group of HA, HA hydrogels can be prepared as well as HA derivatives with amine functional groups. If the ratio is much higher than 1/1, the diamine is grafted as a pendant group.

5.7. Tumor treatment

Existing anticancer drug deliveries focus on self-assembled polymeric nanoparticles based on implantable biomaterials. Due to the special core–shell structure of such nanoparticles, they have superior properties *in vitro* and *in vivo*, including high loading capacity of poorly water-soluble drugs, releasing drugs in a sustained manner, thus increasing bioavailability of them (Hou et al., 2011; Saravanakumar et al., 2010). However, only partial amounts of loaded drugs reach the target site due to some physiological limitations. One approach to overcome these limitations is active targeting strategies such as binding to appropriate receptors highly expressed at the target site. An amphiphilic HA derivative conjugated with glycyrrhetinic acid (GA) was developed to self-assemble nanoparticles for liver tumor targeting delivery of PTX by Zhang, Yao, Zhou, Wang, and Zhang (2013). PTX-loaded HGA nanoparticles exhibited more significant cytotoxicity to HepG2 cells than B16F10 cells due to simultaneously overexpressing HA and GA receptors of HepG2 cells. The targeting activity of HGA nanoparticles was also demonstrated by *in vitro* cellular uptake studies and *in vivo* imaging analysis. Therefore, HGA nanoparticles can be a potential targeting drug carrier for liver cancer therapy.

There is always a close relationship between malignancy and HA content. Higher concentration of HA will be present in the adjacent tissues surrounding invasive tumors than in the corresponding tissues of noninvasive tumors. In some types of tumors, this abnormal increase in the HA level was found to result from the overproduction of HA in fibroblast cells by the interaction with adjacent cancer cells (Asplund, Versnel, Laurent, & Heldin, 1993). Similarly, HA also overproduced in the cancer cells themselves possibly due to overexpression of HA in the cancer cells (Calabro, Oken, Hascall, & Masellis, 2002). Consequently, the high level of HA in the tissues can be a significant predictor for estimating malignancy and invasiveness of tumor (Ropponen et al., 1998). In some specific types of tumors, such as bladder cancer, the urinary level of HA can also be a marker for detecting cancer as well as for evaluating its grade (Lokeshwar et al., 2000). Choi, Saravanakumar, Park, and Park (2012) in their review had concluded that HA-based DDS have great potential for imaging and treatment of several cancers.

PTX, an effective chemotherapeutic drug isolated from the bark of *Taxus brevifolia*, is a microtubule-stabilizing agent that can induce mitotic arrest and cell apoptosis. Currently, PTX is widely used for the treatment of cancer, including ovarian, breast, and non-small-cell lung cancer. Hyaluronic acid-coated, PTX-loaded, nanostructured lipid carriers (HA-NLCs) were prepared via electrostatic attraction for delivering PTX to tumors overexpressing CD44. HA-NLC prepared via electrostatic attraction was an effective carrier for delivering PTX to tumors overexpressing CD44 (Yang et al., 2013).

Hyaluronyl reduced graphene oxide (rGO) nanosheets was used as a tumor targeting delivery system for anticancer agents by Miao et al. (2013). Hyaluronyl-modified rGO nanosheets were prepared by synthesizing cholesteryl hyaluronic acid (CHA) and using it to coat rGO nanosheets, yielding CHA-rGO. Compared with rGO, CHA-rGO nanosheets showed increased colloidal stability under physiological conditions and improved *in vivo* safety, with a survival rate of 100% after intravenous administration of 40 mg/kg in mice. The DOX-loading capacity of CHA-rGO was fourfold greater than that of rGO.

5.8. Environmental applications

A series of DTPA-substituted hyaluronan derivatives with different degrees of substitution and degrees of cross-linking were successfully prepared by Buffa et al. (2011). Several parameters of the reaction such as molecular weight of starting HA, temperature, equivalent of DTPA bis-anhydride, concentration of HA, transacylation catalyst DMAP, and reaction time were studied. A specific benefit of the prepared conjugates of HA–DTPA is that this formulation can effectively complex metals. It could be applied as a metal sponge material (hydrogel), which can effectively remove radioactive or toxic metals *in vivo*.

Natural polysaccharides such as chitosan, heparin, chondroitin, keratin, and xanthan have been developed as environmentally friendly materials for removing toxic pollutants from water. In particular, chitosan derivative of chitin obtained mainly from crab and shrimp shells has been widely suggested as a candidate for an overwhelming scope of adsorption applications, covering almost all the spectrum of biotechnology. HA is another natural polysaccharide that has properties similar to chitosan. Novel magnetic adsorbent with submicrosize was fabricated by Lan et al. (2013) through the immobilization of hyaluronic acid on the magnetic silica microspheres. The as-synthesized Fe_3O_4–SiO_2–HA microspheres can be used as an effective adsorbent for the removal of copper ions from aqueous solution.

5.9. Sensors

Although chitosan is often used to disperse carbon nanotubes (CNTs) with subsequent of electrochemical (bio)sensors (Kac & Ruzgas, 2006), recent studies suggest HA can be efficiently used for the same purpose as judged from excellent conductivity of the nanocomposite and remarkable dispersivity of HA toward CNTs (Razal, Gilmore, & Wallace, 2008). A biocompatible nanocomposite consisting of single-walled CNTs dispersed in a hyaluronic acid (HA) was investigated as a sensing platform for a mediatorless electrochemical detection of NADH by Filip, Sefčovičová, Tomčik, Gemeiner, and Tka (2011). The NADH sensor exhibits a good long-term operational stability (95% of the original sensitivity after 22 h of continuous operation).

6. CONCLUSION

Thus, HA is a ubiquitous natural polysaccharide in the body with excellent physicochemical properties such as biodegradable, biocompatible, nontoxic, and nonimmunogenic characteristics. Accordingly, HA has been widely used for various medical applications such as OA surgery, ocular surgery, plastic surgery, TE, and drug delivery. Especially, due to the great impact of drug delivery applications, HA has been extensively explored as a novel drug carrier for target-specific and long-acting delivery of protein, peptide, and nucleotide therapeutics.

ACKNOWLEDGMENTS

The authors are grateful to authorities of DKM College for Women and Thiruvalluvar University, Vellore, Tamil Nadu, India, for the support. Thanks are also due to the editor Dr. Se-Kwon Kim, Marine Bioprocess Research Center, Pukyong National University, South Korea, for the opportunity to review such an innovating field.

REFERENCES

Ahmadzadeh-Asl, S., Hesaraki, S., & Zamanian, A. (2011). Preparation and characterisation of calcium phosphate-hyaluronic acid nanocomposite bone cement. *Advances in Applied Ceramics*, *110*, 340–345.

Ahmed, T. A. E., & Hincke, M. T. (2010). Strategies for articular cartilage lesion repair and functional restoration. *Tissue Engineering. Part B, Reviews*, *16*(3), 305–329.

Allen, T. M. (2002). Ligand-targeted therapeutics in anticancer therapy. *Nature Reviews. Drug Discovery*, *2*, 750–763.

Amarnath, L. P., Srinivas, A., & Ramamurthi, A. (2006). In vitro hemocompatibility testing of UV-modified hyaluronan hydrogels. *Biomaterials*, *27*, 1416–1424.

Anisha, B. S., Biswas, R., Chennazhi, K. P., & Jayakumar, R. (2013). Chitosan-hyaluronic acid/nano silver composite sponges for drug resistant bacteria infected diabetic wounds. *International Journal of Biological Macromolecules, 62*, 310–320.

Apelt, D., Theiss, F., El-Warrak, A. O., Zlinszky, K., Bettschart-Wolfisberger, R., Bohner, M., et al. (2004). In vivo behavior of three different injectable hydraulic calcium phosphate cements. *Biomaterials, 25*, 1439–1451.

Arnal-Pastor, M., Martínez Ramos, C., Pérez Garnés, M., Monleón Pradas, M., & Vallés Lluch, A. (2013). Electrospun adherent–antiadherent bilayered membranes based on cross-linked hyaluronic acid for advanced tissue engineering applications. *Materials Science and Engineering C, 33*, 4086–4093.

Asplund, T., Versnel, M. A., Laurent, T. C., & Heldin, P. (1993). Human mesothelioma cells produce factors that stimulate the production of hyaluronan by mesothelial cells and fibroblasts. *Cancer Research, 53*(2), 388–392.

Bae, M. S., Ohe, J.-Y., Lee, J. B., Heo, D. N., Byun, W., Bae, H., et al. (2014). Photo-cured hyaluronic acid-based hydrogels containing growth and differentiation factor 5 (GDF-5) for bone tissue regeneration. *Bone, 59*, 189–198.

Balazs, E. A. (2009). Therapeutic use of hyaluronan. *Structural Chemistry, 20*, 341.

Banerji, S., Wright, A. J., Noble, M., Mahoney, D. J., Campbell, I. D., Day, I. D., et al. (2007). Structures of the Cd44-hyaluronan complex provide insight into a fundamental carbohydrate-protein interaction. *Nature Structural & Molecular Biology, 14*, 234–239.

Benjamin Sadowitz, M. D., Keri Seymour, D. O., Vivian Gahtan, M. D., & Maier, K. G. (2012). The role of hyaluronic acid in atherosclerosis and intimal hyperplasia. *Journal of Surgical Research, 173*, e63–e72.

Bohner, M., Gbureck, U., & Barralet, J. E. (2005). Technological issues for the development of more efficient calcium phosphate bone cements: A critical assessment. *Biomaterials, 26*, 6423–6429.

Bot, P. T., Hoefer, I. E., Piek, J. J., & Pasterkamp, G. (2008). Hyaluronic acid: Targeting immune modulatory components of the extracellular matrix in atherosclerosis. *Current Medicinal Chemistry, 15*(8), 786–791.

Boudou, T., Crouzier, T., Ren, K. F., Blin, G., & Picart, C. (2010). Multiple functionalities of polyelectrolyte multilayer films: New biomedical applications. *Advanced Materials, 22*(4), 441–467.

Bryant, S. J., & Anseth, K. S. (2002). Hydrogel properties influence ECM production by chondrocytes photoencapsulated in poly(ethylene glycol) hydrogels. *Journal of Biomedical Materials Research, 59*(1), e63–e72.

Buffa, R., Běták, J., Kettou, S., Hermannová, M., Pospíšilová, L., & Velebný, V. (2011). A novel DTPA cross-linking of hyaluronic acid and metal complexation thereof. *Carbohydrate Research, 346*, 1909–1915.

Burns, J. W., & Valeri, C. R. (1996). Methods for the inhibition of platelet adherence and aggregation. U.S. Patent no. 5,585,361.

Calabro, A., Oken, M. M., Hascall, V. C., & Masellis, A. M. (2002). Characterization of hyaluronan synthase expression and hyaluronan synthesis in bone marrow mesenchymal progenitor cells: Predominant expression of HAS1 mRNA and up-regulated hyaluronan synthesis in bone marrow cells derived from multiple myeloma patients. *Blood, 100*, 2578–2585.

Chang, K.-H., Liao, H.-T., & Chen, J.-P. (2013). Preparation and characterization of gelatin/hyaluronic acid cryogels for adipose tissue engineering: In vitro and in vivo studies. *Acta Biomaterialia, 9*, 9012–9026.

Chen, W. Y., Marcellin, E., Hung, J., & Nielsen, L. K. (2009). Hyaluronan molecular weight is controlled by UDP-N-acetylglucosamine concentration in Streptococcus zooepidemicus. *Journal of Biological Chemistry, 284*, 18007–18014.

Chen, Q., Zhang, S. Z., Ying, H. Z., Dai, X. Y., Li, X. X., Yu, C. H., et al. (2012). Chemical characterization and immunostimulatory effects of a polysaccharide from Polygoni Multiflori Radix Praeparata in cyclophosphamide-induced anemic mice. *Carbohydrate Polymers, 88*, 1476–1482.

Cheung, H. Y., Lau, K. T., Lu, T. P., & Hui, D. (2007). A critical review on polymer-based bio-engineered materials for scaffold development. *Composites Part B: Engineering, 38*(3), 291–300.

Cho, H.-J., Yoon, I.-S., Yoon, H. Y., Koo, H., Jin, Y.-J., Ko, S.-H., et al. (2012). Polyethylene glycol-conjugated hyaluronic acid-ceramide self-assembled nanoparticles for targeted delivery of doxorubicin. *Biomaterials, 33*, 1190–1200.

Choi, J.-i., Kim, J.-K., Kim, J.-H., Kweon, D.-K., & Lee, J.-W. (2010). Degradation of hyaluronic acid powder by electron beam irradiation, gamma ray irradiation, microwave irradiation and thermal treatment: A comparative study. *Carbohydrate Polymers, 79*(4), 1080–1085.

Choi, K. Y., Min, K. H., Yoon, H. Y., Kim, K., Park, J. H., Kwon, I. C., et al. (2011). PEGylation of hyaluronic acid nanoparticles improves tumor targetability in vivo. *Biomaterials, 32*(7), 1880–1889.

Choi, K. Y., Saravanakumar, G., Park, J. H., & Park, K. (2012). Hyaluronic acid-based nanocarriers for intracellular targeting: Interfacial interactions with proteins in cancer. *Colloids and Surfaces B: Biointerfaces, 99*, 82–94.

Chung, C., & Burdick, J. A. (2008). Engineering cartilage tissue. *Advanced Drug Delivery Reviews, 60*(2), 243–262.

Chung, E. J., Jakus, A. E., & Shah, R. N. (2013). In situ forming collagen-hyaluronic acid membrane structures: Mechanism of self-assembly and applications in regenerative medicine. *Acta Biomaterialia, 9*, 5153–5161.

Coessens, V., Pintauer, T., & Matyjaszewski, K. (2001). Functional polymers by atom transfer radical polymerization. *Progress in Polymer Science, 26*, 37–77.

Daamen, W. F., Moerkerk, H. T. B. V., Haffmans, T., Buttafoco, L., Poot, A. A., Veerkamp, J. H., et al. (2003). Preparation and evaluation of molecularly defined collagen–elastin–glycosaminoglycan scaffolds for tissue engineering. *Biomaterials, 24*, 4001–4009.

Dahlmann, J., Krause, A., Möller, L., Kensah, G., Möwes, M., Diekmann, A., et al. (2013). Fully defined in situ cross-linkable alginate and hyaluronic acid hydrogels for myocardial tissue engineering. *Biomaterials, 34*, 940–951.

Deen, I., & Zhitomirsky, I. (2014). Electrophoretic deposition of composite halloysite nanotube–hydroxyapatite–hyaluronic acid films. *Journal of Alloys and Compounds, 586*, S531–S534.

DeNardo, S. J., Yao, Z., Lam, K. S., Song, A., Burke, P. A., Mirick, G. R., et al. (2003). Effect of molecular size of PEGylated peptide on the pharmacokinetics and tumor targeting in lymphoma-bearing mice. *Clinical Cancer Research, 9*, 3854S–3864S.

Diaz-Montero, C. M., Wang, Y., Shao, L. J., Feng, W., Zidan, A. A., Pazoles, C. J., et al. (2012). The glutathione disulfide mimetic NOV-002 inhibits cyclophosphamide-induced hematopoietic and immune suppression by reducing oxidative stress. *Free Radical Biology & Medicine, 52*(9), 1560–1568.

Donnelly, R. F., Majithiya, R., Singh, T. R. R., Morrow, D. I. J., Garland, M. J., Demir, Y. K., et al. (2011). Design, optimization and characterization of polymeric microneedle arrays prepared by a novel laser-based micromoulding technique. *Pharmaceutical Research, 28*, 41–57.

Elder, A. N., Hannes, S. K., Atoyebi, S. F., & Washburn, N. R. (2013). Effects on peptide binding affinity for TNFα by PEGylation and conjugation to hyaluronic acid. *European Polymer Journal, 49*, 2968–2975.

Elisseeff, J., McIntosh, W., Anseth, K., Riley, S., Ragan, P., & Langer, R. (2000). Photoencapsulation of chondrocytes in poly(ethylene oxide)-based semi-interpenetrating networks. *Journal of Biomedical Materials Research, 51*(2), e164–e171.

Fatnassi, M., Tourné-Péteilh, C., Peralta, P., Cacciaguerra, T., Dieudonné, P., Devoisselle, J.-M., et al. (2013). Encapsulation of complementary model drugs in spray-dried nanostructured materials. *Journal of Sol-Gel Science and Technology, 68*(2), 307–316.

Filip, J., Sefčovičová, J., Tomčik, P., Gemeiner, P., & Tka, J. (2011). A hyaluronic acid dispersed carbon nanotube electrode used for a mediatorless NADH sensing and biosensing. *Talanta, 84*, 355–361.

Flautre, B., Lemaitre, J., Maynou, C., Van Landuyt, P., & Hardouin, P. (2003). Influence of polymeric additives on the biological properties of brushite cements: An experimental study in rabbit. *Journal of Biomedical Materials Research, A66*(2), 214–223.

Flückiger, U., & Zimmerli, W. (2000). Factors influencing antimicrobial therapy of surface adhering microorganisms. *Recent Research Development in Antimicrobial Agents and Chemotherapy, 4*, 165–175.

Forsey, R. W., Fisher, J., Thompson, J., Stone, M. H., Bell, C., & Ingham, E. (2006). The effect of hyaluronic acid and phospholipid based lubricants. *Biomaterials, 27*, 4581–4590.

Fraser, J. R. E., Laurent, T. C., & Laurent, U. B. G. (1997). Hyaluronan: Its nature, distribution, functions and turnover. *Journal of Internal Medicine, 242*, 27–33.

Garcia-Abuin, D., Gomez-Diaz, J. M., Navaza, L., & Regueiro, I. V.-T. (2011). Viscosimetric behaviour of hyaluronic acid in different aqueous solutions. *Carbohydrate Polymers, 85*, 500–505.

Geetha, M., Singh, A. K., Asokamani, R., & Gogia, A. K. (2009). Ti based biomaterials, the ultimate choice for orthopaedic implants—A review. *Progress in Materials Science, 54*(3), 397–425.

Giammona, G., Pitarresi, G., Palumbo, F., Romano, L. C., Meani, E., & Cremascoli, E. (2013). Hyaluronic acid based hydrogel and use thereof in surgery. U.S. Patent Application Publication, Publication no. U.S. 2013/0129800A1.

Gilli, P., Ferretti, V., & Gilli, G. (1994). Enthalpy—Entropy compensation in drug receptor binding. *Journal of Physical Chemistry, 98*, 1515–1518.

Glinel, K., Thebault, P., Humblot, V., Pradier, C. M., & Jouenne, T. (2012). Antibacterial surfaces developed from bio-inspired approaches. *Acta Biomaterialia, 8*, 1670–1684.

Hascall, V. C., Calabro, A., Oken, M. M., & Masellis, A. M. (2002). Characterization of hyaluronan synthase expression and hyaluronan synthesis in bone marrow mesenchymal progenitor cells: Predominant expression of HAS1 mRNA and up-regulated hyaluronan synthesis in bone marrow cells derived from multiple myeloma patients. *Blood, 100*(7), 2578–2585.

He, Y., Cheng, G., Li Xie, Y., Nie, B. H., & Zhongwei, G. (2013). Polyethyleneimine/DNA polyplexes with reduction-sensitive hyaluronic acid derivatives shielding for targeted gene delivery. *Biomaterials, 34*(4), 1235–1245.

Heller, J. (1985). Controlled drug release from poly(ortho esters): A surface eroding polymer. *Journal of Controlled Release, 2*, 167–177.

Hilton, J. R., Williams, D. T., Beuker, B., Miller, D. R., & Harding, K. G. (2004). Wound dressings in diabetic foot disease. *Clinical Infectious Diseases, 39*(2), S100–S103.

Hou, L., Fan, Y., Yao, J., Zhou, J. P., Li, C. C., Fang, Z. J., et al. (2011). Low molecular weight heparin-all-trans-retinoid acid conjugate as a drug carrier for combination cancer chemotherapy of paclitaxel and all-trans-retinoid acid. *Carbohydrate Polymers, 86*, 1157–1161.

Huang, S. L., Ling, P. X., & Zhang, T. M. (2007). Oral absorption of hyaluronic acid and phospholipids complexes in rats. *World Journal of Gastroenterology, 13*(6), 945–949.

Hutchison, J. B., Stark, P. F., Hawker, C. J., & Anseth, K. S. (2005). Polymerizable living free radical initiators as a platform to synthesize functional networks. *Chemistry of Materials, 17*(19), 4789–4797.

Ibrahim, S., Kang, Q. K., & Ramamurthi, A. (2010). The impact of hyaluronic acid oligomer content on physical, mechanical, and biologic properties of divinyl sulfone-crosslinked hyaluronic acid hydrogels. *Journal of Biomedical Materials Research. Part A, 94*(2), 355–370.

Ito, Y., Murano, H., Hamasaki, N., Fukushima, K., & Takada, K. (2011). Incidence of low bioavailability of leuprolide acetate after percutaneous administration to rats by dissolving microneedles. *International Journal on Pharmaceuticals, 407*(1–2), 126–131.

Ji, Y., Ghosh, K., Li, B., Sokolov, J. C., Clark, R. A., & Rafailovich, M. H. (2006). Dual-syringe reactive electrospinning of cross-linked hyaluronic acid hydrogel nanofibers for tissue engineering applications. *Macromolecular Bioscience, 6*(10), 811–817.

Ji, Y., Ghosh, K., Shu, X. Z., Li, B., Sokolov, J. C., Prestwich, G. D., et al. (2006). Electrospun three-dimensional hyaluronic acid nanofibrous scaffolds. *Biomaterials, 27*(20), 3782–3792.

Jin, C. Y., Han, M. H., Lee, S. S., & Choi, Y. H. (2009). Mass producible and biocompatible microneedle patch and functional verification of its usefulness for transdermal drug delivery. *Biomedical Microdevices, 11*(6), 1195–1203.

Johns, M. R., Tang Goh, L., & Oeggeru, A. (1994). Effect of pH, agitation and aeration on hyaluronic acid production by Streptococcus zooepidemicus. *Biotechnology Letters, 16*, 507–512.

Kac, J. T., & Ruzgas, T. (2006). Dispersion of single walled carbon nanotubes. Comparison of different dispersing strategies for preparation of modified electrodes toward hydrogen peroxide detection. *Electrochemistry Communications, 8*(5), 899–903.

Ke, C., Sun, L., Qiao, D., Wang, D., & Zeng, X. (2011). Antioxidant activity of low molecular weight hyaluronic acid. *Food and Chemical Toxicology, 49*(10), 2670–2675.

Ke, C., Wang, D., Sun, Y., Qiao, D., Ye, H., & Zeng, X. (2013). Immunostimulatory and antiangiogenic activities of low molecular weight hyaluronic acid. *Food and Chemical Toxicology, 58*, 401–407.

Kim, S. U., & De Vellis, J. (2009). Stem cell-based cell therapy in neurological diseases: A review. *Journal of Neuroscience Research, 87*(10), 2183–2200.

Kim, S. K., Ravichandran, Y. D., Khan, S. B., & Kim, Y. T. (2008). Prospective of the cosmeceuticals derived from marine organisms. *Biotechnology and Bioprocess Engineering, 13*, 511–523.

Kim, J. K., Srinivasan, P., Kim, J. H., Choi, J.-i., Part, H. J., Byun, M. W., et al. (2008). Structural and antioxidant properties of gamma irradiated hyaluronic acid. *Food Chemistry, 109*, 763–770.

Lan, S., Xiaomin, W., Li, L., Li, M., Guo, F., & Gan, S. (2013). Synthesis and characterization of hyaluronic acid-supported magnetic microspheres for copper ions removal. *Colloids and Surfaces A: Physicochemical and Engineering Aspects, 425*, 42–50.

Laurent, T. (1998). *The chemistry, biology and medical applications of hyaluronan and its derivatives*. London: Portland Press.

Lee, F., & Kurisawa, M. (2013). Formation and stability of interpenetrating polymer network hydrogels consisting of fibrin and hyaluronic acid for tissue engineering. *Acta Biomaterialia, 9*, 5143–5152.

Lequeux, I., Ducasse, E., Jouenne, T., & Thebault, P. (2014). Addition of antimicrobial properties to hyaluronic acid by grafting of antimicrobial peptide. *European Polymer Journal, 51*, 182–190.

Levett, P. A., Ferry, P. W., Melchels, K. S., Hutmacher, D. W., Malda, J., & Klein, T. J. (2014). A biomimetic extracellular matrix for cartilage tissue engineering centered on photocurable gelatin, hyaluronic acid and chondroitin sulfate. *Acta Biomaterialia, 10*, 214–223.

Li, Q., Niinomi, M., Nakai, M., Cui, Z. D., Zhu, S. L., & Yang, X. J. (2011). Improvements in the super elasticity and change in deformation mode of beta-type TiNb(24)Zr(2) alloys caused by aging treatments. *Metallurgical and Materials Transactions A, 42A*, 2843–2849.

Linhardt, R. J. (2001). Analysis of glycosaminoglycans with polysaccharidelyases. *Current Protocol Molecular Biology, 17*, 13B.

Liu, L., Du, G., Chen, J., Wang, M., & Sun, J. (2008). Enhanced hyaluronic acid production by a two-stage culture strategy based on the modeling of batch and fed-batch cultivation of Streptococcus zooepidemicus. *Bioresource Technology, 99*, 8532–8536.

Liu, S., Jin, M.-N., Quan, Y.-S., Kamiyama, F., Kusamori, K., Katsumi, H., et al. (2014). Transdermal delivery of relatively high molecular weight drugs using novel self-dissolving microneedle arrays fabricated from hyaluronic acid and their characteristics and safety after application to the skin. *European Journal of Pharmaceutics and Biopharmaceutics, 86*, 267–276. http://dx.doi.org/10.1016/j.ejpb.2013.10.001.

Liu, L., Liu, D., Wang, M., Du, G., & Chen, J. (2007). Preparation and characterization of sponge-like composites by cross-linking hyaluronic acid and carboxymethylcellulose sodium with adipic dihydrazide. *European Polymer Journal, 43*, 2672–2681.

Liu, Y., Ma, G., Fang, D., Xu, J., Zhang, H., & Nie, J. (2011). Effects of solution properties and electric field on the electrospinning of hyaluronic acid. *Carbohydrate Polymers, 83*(2), 1011–1015.

Lokeshwar, V. B., Ek, C. A. N., Pham, H. T., Wei, D., Young, M. J., Duncan, R. C., et al. (2000). Urinary hyaluronic acid and hyaluronidase: Markers for bladder cancer detection and evaluation of grade. *Journal of Urology, 163*, 348–356.

Lumelsky, N., Blondel, O., Laeng, P., Velasco, I., Ravin, R., & McKay, R. (2001). Differentiation of embryonic stem cells to insulin-secreting structures similar to pancreatic islets. *Science, 292*(5520), 1389–1394.

Luo, Y., Kirker, K., & Prestwich, G. (2000). Cross-linked hyaluronic acid hydrogel films: New biomaterials for drug delivery. *Journal of Controlled Release, 69*, 169–184.

Ma, Z., Gao, C., Gong, Y., & Shen, J. (2005). Cartilage tissue engineering PLLA scaffold with surface immobilized collagen and basic fibroblast growth factor. *Biomaterials, 26*(11), 1253–1259.

Macknight, W. J., Ponomarenko, E. A., & Tirrell, D. A. (1998). Self-assembled complexes in nonaqueous solvents and in the solid state. *Accounts of Chemical Research, 31*, 781–788.

Mah, T. F. C., & O'Toole, G. A. (2001). Mechanisms of biofilm resistance to antimicrobial agents. *Trends in Microbiology, 9*, 34–39.

Masaoka, Y., Tanaka, Y., Kataoka, M., Sakuma, S., & Yamashita, S. (2006). Site of drug absorption after oral administration: Assessment of membrane permeability and luminal concentration of drugs in each segment of gastrointestinal tract. *European Journal of Pharmaceutical Sciences, 29*(3–4), 240–250.

Miao, W., Shim, G., Kang, C. M., Lee, S., Choe, Y. S., Choi, H.-G., et al. (2013). Cholesteryl hyaluronic acid-coated, reduced graphene oxide nanosheets for anti-cancer drug delivery. *Biomaterials, 34*, 9638–9647.

Nageeb, M., Nouh, S. R., Bergman, K., Nagy, N. B., Khamis, D., Kisiel, M., et al. (2012). Bone engineering by biomimetic injectable hydrogel. *Molecular Crystals and Liquid Crystals, 555*, 177–188.

Nair, S., Remya, N. S., Remya, S., & Nair, P. D. (2011). A biodegradable in situ injectable hydrogel based on chitosan and oxidized hyaluronic acid for tissue engineering applications. *Carbohydrate Polymers, 85*, 838–844.

Oliveira, J. M., Rodriguesa, M. T., Silva, S. S., Malafaya, P. B., Gomes, M. E., Viegas, C. A., et al. (2006). Novel hydroxyapatite/chitosan bilayered scaffold for osteochondral tissue-engineering applications: Scaffold design and its performance when seeded with goat bone marrow stromal cells. *Biomaterials, 27,* 6123–6137.

Ouasti, S., Donno, R., Cellesi, F., Sherratt, M. J., Terenghi, G., & Tirelli, N. (2011). Network connectivity, mechanical properties and cell adhesion for hyaluronic acid/PEG hydrogels. *Biomaterials, 32,* 6456–6470.

Oyarzun-Ampuero, F. A., Goycoolea, F. M., Torres, D., & Alonso, M. J. (2011). A new drug nano carrier consisting of polyarginine and hyaluronic acid. *European Journal of Pharmaceutics and Biopharmaceutics, 79*(1), 54–57.

Palumbo, F. S., Pitarresi, G., Fiorica, C., Rigogliuso, S., Ghersi, G., & Giammon, G. (2013). Chemical hydrogels based on a hyaluronic acid-graft-α-elastin derivative as potential scaffolds for tissue engineering. *Materials Science and Engineering C, 33,* 2541–2549.

Park, S., Bhang, S. H., La, W.-G., Seo, J., Kim, B.-S., & Char, K. (2012). Dual roles of hyaluronic acids in multilayer films capturing nanocarriers for drug-eluting coatings. *Biomaterials, 33*(21), 5468–5477.

Pescosolido, L., Vermonden, T., Malda, J., Censi, R., Dhert, W. J., Alhaique, F., et al. (2011). In situ forming IPN hydrogels of calcium alginate and dextran-HEMA for biomedical applications. *Acta Biomaterialia, 7*(4), 1627–1633.

Piao, M. G., Kim, J. H., Kim, J. O., Lyoo, W. S., Lee, M. H., Yong, C. S., et al. (2007). Enhanced oral bioavailability of piroxicam in rats by hyaluronate microspheres. *Drug Development and Industrial Pharmacy, 33*(4), 485–491.

Picotti, F., Fabbian, M., Gianni, R., Sechi, A., Stucchi, L., & Bosco, M. (2013). Hyaluronic acid lipoate: Synthesis and physicochemical properties. *Carbohydrate Polymers, 93,* 273–278.

Pitarresi, G., Palumbo, F. S., Calascibetta, F., Fiorica, C., Di Stefano, M., & Giammona, G. (2013). Medicated hydrogels of hyaluronic acid derivatives for use in orthopedic field. *International Journal of Pharmaceutics, 449,* 84–94.

Puré, E., & Assoian, R. K. (2009). Rheostatic signaling by CD44 and hyaluronan. *Cellular Signalling, 21*(5), 651–655.

Razal, J. M., Gilmore, K. J., & Wallace, G. G. (2008). Carbon nanotube biofiber formation in a polymer-free coagulation bath. *Advanced Functional Materials, 18*(1), 61–66.

Romagnoli, M., & Belmontesi, M. (2008). Hyaluronic acid-based fillers: Theory and practice. *Clinical Dermatology, 26*(2), 123–159.

Romagnoli, E., Mascia, M. L., Cipriani, C., Fassino, V., Mazzei, F., & D'Erasmo, E. (2008). Short and long-term variations in serum calciotropic hormones after a single very large dose of ergocalciferol (vitamin D2) or cholecalciferol (Vitamin D3) in the elderly. *The Journal of Clinical Endocrinology and Metabolism, 93,* 3015–3020.

Ropponen, K., Tammi, M., Parkkinen, J., Eskelinen, M., Tammi, R., Lipponen, P., et al. (1998). Tumor cell-associated hyaluronan as an unfavorable prognostic factor in colorectal cancer. *Cancer Research, 58*(2), 342–347.

Saravanakumar, G., Choi, K. Y., Yoon, H. Y., Kim, K., Park, J. H., Kwon, I. C., et al. (2010). Hydrotropic hyaluronic acid conjugates: Synthesis, characterization, and implications as a carrier of paclitaxel. *International Journal of Pharmaceutics, 394,* 154–161.

Sasisekharan, R., Raman, R., & Prabhakar, V. (2006). Glycomics approach to structure-function relationships of glycosaminoglycans. *Annual Review of Biomedical Engineering, 8,* 181–231.

Schanté, C. E., Zuber, G., Herlin, C., & Vandamme, T. F. (2012). Improvement of hyaluronic acid enzymatic stability by the grafting of amino-acids. *Carbohydrate Polymers, 87,* 2211–2216.

Schwartz, J. A., Lantis, J. C., Gendics, C., Fuller, A. M., Payne, W., & Ochs, D. (2013). A prospective, non comparative, multicenter study to investigate the effect of cadexomer iodine on bioburden load and other wound characteristics in diabetic foot ulcers. *International Wound Journal, 10*(2), 193–199.

Seidlits, S. K., Drinnan, C. T., Petersen, R. R., Shear, J. B., Suggs, L. J., & Schmidt, C. E. (2011). Fibronectin-hyaluronic acid composites for three-dimensional endothelial cell culture. *Acta Biomaterialia, 7*, 2401–2409.

Shah, M. V., Badle, S. S., & Ramachandran, K. B. (2013). Hyaluronic acid production and molecular weight improvement by redirection of carbon flux towards its biosynthesis pathway. *Biochemical Engineering Journal, 80*(2013), 53–60.

Shiedlin, A., Bigelow, R., Christopher, W., Arbabi, S., Yang, L., Maier, R. V., et al. (2004). Evaluation of hyaluronan from different sources: Streptococcus zooepidemicus, rooster comb, bovine vitreous and human umbilical cord. *Biomacromolecules, 5*, 2122–2127.

Son, Y.-J., Yoon, I.-S., Sung, J.-H., Cho, H.-J., Chung, S.-J., Shim, C.-K., et al. (2013). Porous hyaluronic acid/sodium alginate composite scaffolds for human adipose-derived stem cells delivery. *International Journal of Biological Macromolecules, 61*, 175–181.

Souto, E. B., & Muller, R. H. (2008). Cosmetic features and applications of lipid nanoparticles (SLN®, NLC®). *International Journal of Cosmetic Science, 30*, 157–165.

Stern, R. (2003). Devising a pathway for hyaluronan catabolism: Are we there yet? *Glycobiology, 13*(12), 105R–115R.

Tan, H., Ramirez, C. M., Miljkovic, N., Han Li, J., Rubin, P., & Marra, K. G. (2009). Thermosensitive injectable hyaluronic acid hydrogel for adipose tissue engineering. *Biomaterials, 30*, 6844–6853.

Tolentino, A., Alla, A., de Ilarduya, A. M., & Guerra, S. M. (2013). Comb-like ionic complexes of hyaluronic acid with alkyltrimethylammonium Surfactants. *Carbohydrate Polymers, 92*, 691–696.

Toole, B. P., Wight, N. T., & Tammi, I. M. (2001). Hyaluronan–cell interactions in cancer and vascular disease. *Journal of Biological Chemistry, 277*, 4593–4596.

Turner, N. J., Kielty, C. M., Walker, M. G., & Canfield, A. E. (2004). A novel hyaluronan-based biomaterial (Hyaff-11) as a scaffold for endothelial cells in tissue engineered vascular grafts. *Biomaterials, 25*, 5955–5964.

Vasiliu, S., Popa, M., & Rinaudo, M. (2005). Polyelectrolyte capsules made of two biocompatible natural polymers. *European Polymer Journal, 41*, 923–932.

Vázquez, J. A., Montemayor, M. I., Fraguas, J., & Murado, M. A. (2009). High production of hyaluronic and lactic acids by Streptococcus zooepidemicus in fed-batch culture using commercial and marine peptones from fishing by-products. *Biochemical Engineering Journal, 44*, 125–130.

Wang, T.-W., & Spector, M. (2009). Development of hyaluronic acid-based scaffolds for brain tissue engineering. *Acta Biomaterialia, 5*(7), 2371–2381.

Wang, X., Um, I. C., Fang, D., Okamoto, A., Hsiao, B. S., & Chu, B. (2005). Formation of water-resistant hyaluronic acid nanofibers by blowing-assisted electro-spinning and nontoxic post treatments. *Polymer, 46*(13), 4853–4867.

Wang, H., Wang, M. Y., Chen, J., Tang, Y., Dou, J., Yu, J., et al. (2011). A polysaccharide from Strongylocentrotus nudus eggs protects against myelosuppression and immunosuppression in cyclophosphamide-treated mice. *International Immunopharmacology, 11*, 1946–1953.

William, J. J., & Keith, G. H. (2003). Diabetic foot ulcers. *Lancet, 361*, 1545–1551.

Wohlrab, W., Neubert, R., & Wohlrab, J. (2004). Hyaluronsäure und Haut. In J. Wohlrab (Ed.), *Trends in clinical and experimental dermatology* (pp. 5–39). Aachen: Shaker Verlag.

Xu, X., Jha, A. K., Harrington, D. A., Farach-Carson, M. C., & Jia, X. Q. (2012). Hyaluronic acid-based hydrogels: From a natural polysaccharide to complex networks. *Soft Matter, 8*(12), 3280–3294.

Xu, S., Li, J., He, A., Liu, W., Jiang, X., Zheng, J., et al. (2009). Chemical crosslinking and biophysical properties of electrospun hyaluronic acid based ultra-thin fibrous membranes. *Polymer, 50*(15), 3762–3769.

Yadav, A. K., Mishra, P., & Agrawal, G. P. (2008). An insight on hyaluronic acid in drug targeting and drug delivery. *Journal of Drug Targeting*, *16*(2), 91–107.

Yamada, T., & Kawasaki, T. (2005). Microbial synthesis of hyaluronan and chitin: New approaches. *Journal of Bioscience and Bioengineering*, *99*, 521–528.

Yamanlar, S., Sant, S., Boudou, T., Picart, C., & Khademhosseini, A. (2011). Surface functionalization of hyaluronic acid hydrogels by polyelectrolyte multilayer films. *Biomaterials*, *32*, 5590–5599.

Yan, S., Zhang, Q., Jiannan Wang, Y., Liu, S. L., Li, M., & Kaplan, D. L. (2013). Silk fibroin/chondroitin sulfate/hyaluronic acid ternary scaffolds for dermal tissue reconstruction. *Acta Biomaterialia*, *9*, 6771–6782.

Yang, X.-Y., Li, Y.-X., Li, M., Zhang, L., Feng, L.-X., & Zhang, N. (2013). Hyaluronic acid-coated nanostructured lipid carriers for targeting paclitaxel to cancer. *Cancer Letters*, *334*, 338–345.

Yeap, J. S., Lim, J. W., Vergis, M., Au Yeung, P. S., Chiu, C. K., & Singh, H. (2006). Prophylactic antibiotics in orthopaedic surgery, guidelines and practice. *Medical Journal of Malaysia*, *61*, 181–188.

Yelin, E. (1992). The economics of osteoarthritis. In K. D. Brandt, M. Doherty, & L. S. Lohmander (Eds.), *Osteoarthritis* (pp. 23–30). New York: Oxford Medical Publications.

Yue, W. (2012). Preparation of low-molecular-weight hyaluronic acid by ozone treatment. *Carbohydrate Polymers*, *89*, 709–712.

Wang, J.-S., & Matyjaszewski, K. (1995). Controlled/"living" radical polymerization. Atom transfer radical polymerization in the presence of transition-metal complexes. *Journal of the American Chemical Society*, *117*, 5614–5615.

Zhang, R., Huang, Z. B., Xue, M. Y., Yang, J., & Tan, T. W. (2011). Detailed characterization of an injectable hyaluronic acid-polyaspartylhydrazide hydrogel for protein delivery. *Carbohydrate Polymers*, *85*(4), 717–725.

Zhang, X., Li, Z., Yuan, X., Cuia, Z., & Yang, X. (2013). Fabrication of dopamine-modified hyaluronic acid/chitosan multilayers on titanium alloy by layer-by-layer self-assembly for promoting osteoblast growth. *Applied Surface Science*, *284*, 732–737.

Zhang, L., Yao, J., Zhou, J., Wang, T., & Zhang, Q. (2013). Glycyrrhetinic acid-graft-hyaluronic acid conjugate as a carrier for synergistic targeted delivery of antitumor drugs. *International Journal of Pharmaceutics*, *441*, 654–664.

Zhitomirsky, I. (2002). Fundamental aspects. *Advances in Colloid and Interface Science*, *97*, 277–315.

Zhu, C., Fana, D., & Wang, Y. (2014). Human-like collagen/hyaluronic acid 3D scaffolds for vascular tissue engineering. *Materials Science and Engineering C*, *34*, 393–401.

Ziedler, H. (1986). Synovialflüssigkeit. In W.-M. Kulicke (Ed.), *Fliebverhalten von Stoffen und Stoffgemischen* (pp. 405–433). Basal: Hüthig und Wepf.

CHAPTER TEN

Fucoidans from Marine Algae as Potential Matrix Metalloproteinase Inhibitors

Noel Vinay Thomas*, Se-Kwon Kim[†,1]

*Marine Biochemistry Laboratory, Department of Chemistry, Pukyong National University, Busan, South Korea
[†]Department of Marine-bio Convergence Science, Specialized Graduate School Science and Technology Convergence, Marine Bioprocess Research Center, Pukyong National University, Busan, South Korea
[1]Corresponding author: e-mail address: sknkim@pknu.ac.kr

Contents

1. Introduction	178
2. Sulfated Polysaccharides as Potential MMPIs	182
2.1 Galactans as potential MMPIs	182
2.2 Fucoidans as potential MMPIs	184
3. Conclusions and Further Prospects	188
Acknowledgments	189
References	189

Abstract

Matrix metalloproteinases are endopeptidases which belong to the group of metalloproteinases that contribute for the extracellular matrix degradation and several tissue remodeling processes. An imbalance in the regulation of these endopeptidases eventually leads to several severe pathological complications like cancers, cardiac, cartilage, and neurological-related diseases. Hence, inhibitory substances of metalloproteinases (MMPIs) could prove beneficial in the management of above specified pathological conditions. The available synthetic MMPIs that have been reported until now have few shortcomings, and thus many of them could not make to the final clinical trials. Hence, a growing interest among researchers on screening of MMPIs from different natural resources is evident and especially natural products from marine origin. As there has been an unparalleled contribution of several biologically active compounds from marine resources that have shown a profound applications in nutraceuticals, cosmeceuticals, and pharmaceuticals, we have attempted to discuss the various MMPIs from edible seaweeds.

1. INTRODUCTION

Matrix metalloproteinases (MMPs) are zinc-dependent endopeptidases and were first described almost five decades ago (Gross & Lapiere, 1962). They are known to play a crucial role in various physiological processes that include tissue remodeling and organ development (Page-McCaw, Ewald, & Werb, 2007). MMPs degrade the extracellular matrix which is general remodeling process that enables several pathologic conditions including inflammatory, vascular, and autoimmune disorders, and carcinogenesis as well as physiological processes like wound healing, bone resorption, uterine involution, and organogenesis (Egeblad & Werb, 2002; Lee & Murphy, 2004). These endopeptidases are made up of enzymes namely serralysins, astacins, adamalysins (a disintegrin and metalloproteinase domain or ADAMs), and matrixins (MMPs) which are from metzincin family (Gill & Parks, 2008). There are more than 20 enzymes that are broadly classified into six groups as shown in Fig. 10.1 (Velinov, Poptodorov, Gabrovski, & Gabrovski, 2010).

MMPs are comprised of different subdomains. All MMPs have a common "minimal domain" that is made up of three principal regions. Amino-terminal signals sequence (Pre), a prodomain (Pro) containing a thiol group (SH), and furin (Mayer, Rodríguez, Berlinck, & Fusetani, 2011) cleavage site, and zinc-binding (Zn^{2+}) catalytic domain. The thiol group's interaction with the zinc ion in the catalytic domain keeps the MMP as inactive zymogen. For the activation of zymogen, furin-like proteinases target the furin recognition motif (Mayer et al., 2011) that is found in between prodomain

Enzyme group	MMP type	Reference
Collagenases	MMP-1, -8, -13, and -18	
Gelatinases	MMP-2 and -9	
Stromelysins	MMP-3, -7, -10, -11, -26, -27	[6]
Elastases	MMP-12	
Membrane-type-specific MMPs	MMP-14, -15, -16, -17, -24 H -25	
Other MMPs	MMP-19, -20, -28, -21, -22, -23	

Figure 10.1 Different classes of reported MMPs.

and the catalytic domain. In addition, majority of MMPs have a hemopexin-like region which is composed of four repeats of hemopexin resembling units and contain a disulfide bond (S—S) between the first and the fourth domain. This hemopexin region is connected to the catalytic domain by a flexible hinge region. Apart from the differential structural domains, MMPs are divided into secreted MMPs like MMP-1, -2, -3, -7, -8, -9, -10, -11, -12, -13, -19, -20, -21, -22, -27, -28 and membrane-anchored proteinases like MMP-14, -15, -16, -17, -23, -24, -25, the latter of which use either a transmembrane domain, TM with a cytoplasmic domain, CY attached to it or a glycosylphosphophatidylinositol, GPI anchor, or an amino-terminal signal anchor, SA, which is only in the case of MMP-23, because it is anchored in the plasma membrane. In the case of MMP-23, a unique cysteine array, CA and an immunoglobulin, Ig-like domain is also found. Repeats of collagen-binding type II motif of fibronectin, FN are found in the case of MMP-2 and -9 (Kessenbrock, Plaks, & Werb, 2010) as shown in Fig. 10.2.

MMPs are considered as ideal targets for the treatment of various kinds of cancers because of their apparent role in the pathological processes like apoptosis, signal transduction, transformation of cancer cells, metastasis, tumor invasion, and angiogenesis (Gupta, Kim, Prasad, & Aggarwal, 2010; Overall & López-Otín, 2002). The perilous role of MMPs in the turnover of connective tissue has gained much importance during the past three decades due to the commitment of various research groups across the globe to unwind the mechanisms followed by MMPs in various kinds of human diseases (Malemud, 2006). The intercellular regulation and cell matrix adhesion are generally operated in an organized way; however, majority of cancers that are related to humans are because of uncontrolled regulation of these two phenomena. These pathological changes are because of the

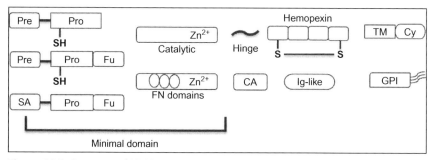

Figure 10.2 Structure of MMPs.

overexpressions of the proteolytic enzymes, MMPs (Bourboulia & Stetler-Stevenson, 2010). It has been reported that matrix degradation is neither the sole nor the main function of these proteases. Indeed, new findings suggest that MMPs act on the proinflammatory cytokines, chemokines, and other proteins and regulate various aspects of inflammatory and immune responses (Parks, Wilson, & López-Boado, 2004).

MMPs are synthesized as prozymogens and are either membrane-bound or secreted and subsequently processed and activated to exhibit the substrate degradation activity. In the process of MMP regulation, cell surface proteins play a key role in activating the MMP proteins (Brooks et al., 1996; Yu & Stamenkovic, 2000). And on the other hand, tissue inhibitors of metalloproteinases (TIMPs) are naturally occurring inhibitors to control unwanted superior expression of MMPs and are known to prevent the proteolytic degradation. Factors like solubility and interaction of TIMPs with Pro-MMPs does determine the inhibitory activity of TIMPs (Lambert, Dassé, Haye, & Petitfrère, 2004; Olson, Gervasi, Mobashery, & Fridman, 1997). During the past 15 years, there is a steady increase of scientific reports confirming the role of MMPs and TIMPs in the tumor microenvironment. For instance, it was reported that uncontrolled expression of MMP-3 in normal breast epithelium results in the formation of invasive tumor by prompting epithelial-to-mesenchymal transition and increased genomic instability (Egeblad & Werb, 2002). It was also observed that MMPs alter the cell–cell adhesion by cleaving E-cadherin that results in superior cell motility, which is an outcome of the epithelial-to-mesenchymal transition. In addition, activation of growth factor signaling by MMPs results in the increased bioavailability of tumor progression factors, such as transforming growth factor-β, fibroblast growth factor (FGF)-2, and vascular endothelial growth factor (VEGF)-A, due to which tumor fibroblast stimulation and angiogenesis result (Shuman Moss, Jensen-Taubman, & Stetler-Stevenson, 2012). Hence, there is a steady increase in the interest of studies on the role of metalloproteinases and their biological importance.

Several pathological responses associated with MMPs have made them as therapeutic targets for the better management of human cancers. Three decades ago, the management of cancer by broadly targeting collagenase (MMP-1), stromelysin (MMP-3), and gelatinase A (MMP-2) was found on reducing degradation of basement membrane and the ECM proteins by cancer cells in metastasis and angiogenesis (Hodgson, 1995; Liotta et al., 1980). Synthetic drugs like Batimastat (BB-94) and Marimastat (BB-2516) have been successful in lowering the expression of MMPs.

However, improper metabolism, low oral bioavailability, poor solubility, side effects like musculoskeletal pain and inflammation, complications, and the risk of increased drug toxicity are still a big challenge (Coussens, Fingleton, & Matrisian, 2002; Thomas & Kim, 2010). Because of these shortcomings of the synthetic MMPIs, researchers are screening natural resources for the screening of MMPIs. And few research groups have been successful in reporting natural MMPIs from terrestrial organisms (Ha et al., 2004; Seo et al., 2005). However, the diversified environment in sea water enables the marine organisms with unique abilities for their survival and makes them ideal choice for the screening of biologically active compounds. Few commercially successful synthetic MMPIs are shown in Fig. 10.3.

Marine macro algae have become a part of oriental diet and also as alternative medicine in Asian-pacific region (Ali et al., 2000), and in the western countries, they are exclusively used for the extraction of bioactives that could be used in pharmaceutical, cosmetics, and food industries (Gómez-Ordóñez, Jiménez-Escrig, & Rupérez, 2010). In recent times, isolation and characterization of the biologically important compounds has gained lot of attention from various research groups across the world. Marine brown algae have been extensively studied for their biologically active components that majorly include polyphenolic derivatives called phlorotannins and polysaccharides like fucoidans and alginic acid (Thomas & Kim, 2013). In this review, we would be discussing the role of phlorotannins and polysaccharides from marine macro algae as potential MMPIs.

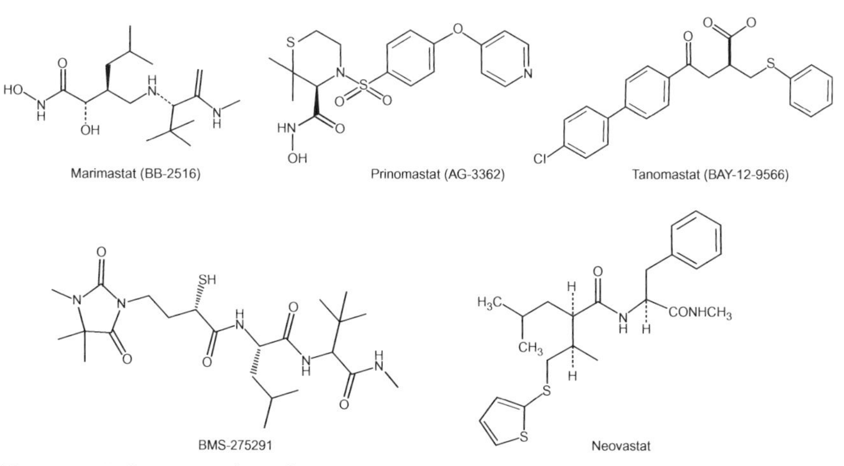

Figure 10.3 Commercial synthetic MMPIs.

2. SULFATED POLYSACCHARIDES AS POTENTIAL MMPIs

Sulfated polysaccharides (SPs) are complex group of sugars with a variety of important biological properties. These polymers which are anionic in nature are widespread in nature and occur in organisms such as mammals and invertebrates (Mourão, 2007; Pomin & Mourão, 2008). Apart from terrestrial organisms and plants, algae from marine sources are considered to be the most abundant resources for the nonanimal SPs. It has been reported that the structure of the SPs is different from species to species (Li, Lu, Wei, & Zhao, 2008; Mayer et al., 2011). For instance, in marine algae, SPs are found in varying amounts among the classes Rhodophyta, Phaeophyta, and Chlorophyta. SPs found in Rhodophyta are called galactans which are comprised of galactose or modified galactose units (Mayer et al., 2011; Pomin & Mourão, 2008). The class of Phaeophyta comprises families of polydisperse molecules based on sulfated L-fucose which are termed as fucans (Berteau & Mulloy, 2003; Bilan & Usov, 2008; Usov & Bilan, 2009). The polydisperse heteropolysaccharides along with traces of homopolysaccharides are the major polysaccharides found in Chlorophyta (Costa et al., 2010; Farias et al., 2008; Hayakawa et al., 2000; Matsubara et al., 2001). SPs exhibit wide range of biological activities such as anticoagulant, antioxidant, antiproliferative, antitumoral, anticomplementary, anti-inflammatory, antiviral, antipeptic, and antiadhesive activities (Cumashi et al., 2007; Damonte, Matulewicz, & Cerezo, 2004; de Azevedo et al., 2009; Ghosh et al., 2009). The molecular size of the SPs is one of the important factors for their effective biological activities (Albuquerque et al., 2004; Silva et al., 2005). In addition, some structural factors like sulfate clusters which guarantees the interaction of SPs with cationic proteins are required for their effective biological activities (Mulloy, 2005).

2.1. Galactans as potential MMPIs

Galactans are SPs found in red seaweeds and are commercially important due to their extensive usage as gelling and thickening agents in the food industry because of their rheological properties. These galactans based on their stereochemistry are basically classified as agarans and carrageenans. Agarans are galactans with 4-linked α-galactose residues of the L-series and carrageneens are galactans with the D-series (Knutsen, Myslabodski, Larsen, & Usov, 1994). These SPs are exclusively found in marine red algae and are extensively used in food industries as emulsifiers, stabilizers, or thickeners. The class of carrageenans consists of a repeating disaccharide backbone of alternating 3-linked β-D-galactopyranose (G) and 4-linked α-D-galactopyranose (D), with 3,6-anhydrogalactose residues commonly present. Based on the number and position of anionic O-sulfo (sulfate) groups, many types of carrageenans have been recognized. Out of them, the most commercially exploited carrageenans are kappa- (κ, DA-G4S), iota- (ι, DA2S-G4S), and lambda- (λ, D2S, 6S-G2S) carrageenans, differ by the presence of one, two, and three sulfate ester groups per repeating disaccharide unit, respectively (McLean & Williamson, 1979; Santos, 1989; Yuan et al., 2005). The carrageenans have been reported to exhibit anticoagulant activity (Farias, Nazareth, & Mourão, 2001; Opoku, Qiu, & Doctor, 2006), antiviral activity (Chiu, Chan, Tsai, Li, & Wu, 2012; Girond, Crance, Van Cuyck-Gandre, Renaudet, & Deloince, 1991; Talarico et al., 2005; Wang et al., 2012), and antitumor activity (Hoffman, Burns, & Paper, 1995). The chemical structure unit of carrageenan is shown in Fig. 10.4.

Angiogenesis is a process where new blood vessels are formed from the existing ones and involves degradation of vascular basement membrane and remodeling of the ECM with the aid of MMPs. MMPs can also negatively contribute to angiogenesis (Rundhaug, 2005) and could promote cancer and cardiac-related diseases (Zhang & Kim, 2009). A study on the anti-angiogenesis effects of low molecular weight and highly sulfated λ-carrageenan oligosaccharides (λ-CO) obtained by depolymerization suggests that λ-CO has effectively inhibited the angiogenesis in the chick chorioallantoic membrane, CAM model and human umbilical vein endothelial cells. The observations show that there was a significant inhibition of the vessel growth at a concentration of 200 μg/pellet. Results of the

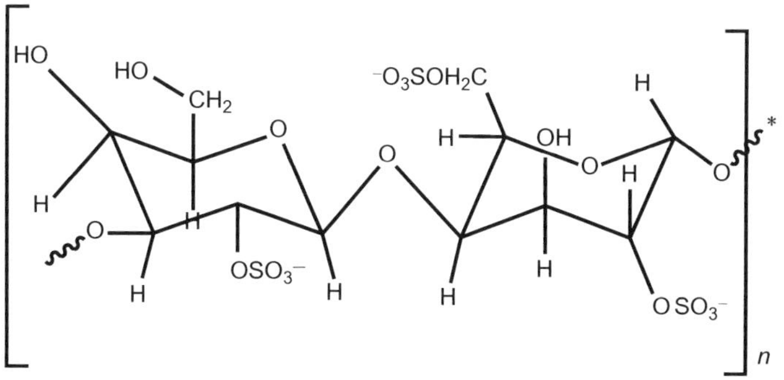

Figure 10.4 Chemical structure unit of carrageenan.

histochemical assays revealed that there is a decrease in the capillary plexus and connective tissues especially in the λ-CO-treated samples. Furthermore, it was also observed that the λ-CO has successfully inhibited the endothelial cell invasion and migration at concentrations of 150–300 μg/ml by the downregulation of MMP-2 expression on endothelial cells (Chen, Yan, Lin, Wang, & Xu, 2007).

Agarans are SPs that are exclusively found in marine red algae. An agaran-type polysaccharide, GFP08 was isolated from *Grateloupia filicina* and was evaluated for its antiangiogenic activity. Based on the CMA assay results, it was reported that GFP08 has reduced the new vessel formation in a dose-dependent manner. In addition, GFP08 has also inhibited the differentiation of HUVECs and also reduced the number of migrated cells (Yu et al., 2012). A chemical structure unit of agaran is shown in Fig. 10.5.

2.2. Fucoidans as potential MMPIs

Fucoidan, a complex SP, is exclusively found in the cell-wall matrix of marine brown algae, some terrestrial plants, animals, and microorganisms (Smestad Paulsen, 2002; Yang & Zhang, 2009). Fucoidans are made up of polymeric carbohydrate structures that are composed of various monosaccharide units linked by glucosidic bonds (Holdt & Kraan, 2011). Fucoidan is composed of a polymer of $\alpha 1 \rightarrow 3$-linked 1-fucose with sulfate groups substituted at the fourth positions some of the fucose residues (Patankar, Oehninger, Barnett, Williams, & Clark, 1993). A chemical structure of fucoidan unit is shown in Fig. 10.6. Recently, fucoidan is being studied extensively due to potential antitumor, antiviral, anticomplement, and anti-inflammatory activities (Chizhov et al., 1999). The low-molecular-weight fucoidan (LMWF)

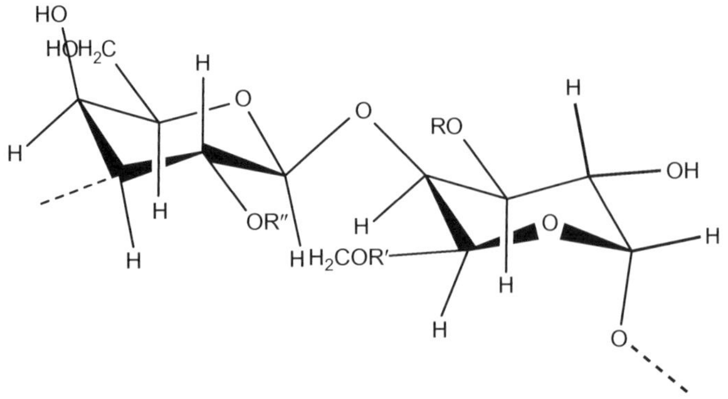

Figure 10.5 Chemical structure unit of agaran.

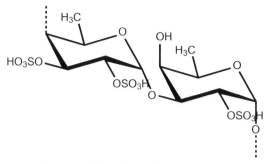

Figure 10.6 Chemical structure of fucoidan unit.

from brown algae and their therapeutic potential on vascular smooth muscle cell (VSMC) and human vascular endothelial cell (HUVECs) proliferation and migration *in vitro* and in a rat model of intimal hyperplasia were investigated. The balloon injury in the thoracic aorta was made to Sprague–Dawley rats which were followed by treatment with either LMWF (5 mg/kg/day) or vehicle for 2 weeks. At the 14th day of morphological analysis and proliferating cell nuclear antigen immunostaining, this investigation revealed that the intimal hyperplasia in rat thoracic aorta was effectively prevented by LMWF when compared with vehicle (neointima area, 3 ± 0.50 mm^2 vs. 5 ± 0.30 mm^2, $P<0.01$). Moreover, significant decrease of MMP-2 expression in the neointima by LMWF was confirmed by the *in situ* zymography. In the same investigation, *in vitro* studies have reported that there was a $45 \pm 5\%$ increase in the migration of HUVECs and $40 \pm 3\%$ decrease in the migration of VSMC by the treatment of LMWF at a concentration of 10 µg/ml. It was suggested that there was increase in migration in HUVECs due to the increase in the MMP-2 mRNA expression by 5.4-fold at 10 µg/ml of LMWF whereas it was 6-fold lower in VSMCs than in untreated control cells (Hlawaty et al., 2011).

Fucoidan obtained from *Fucus vesiculosus*, marine algae, has been reported to show antimetastatic effects in A549 lung cancer cells. This investigation reports that fucoidan downregulated the ERK1/2 and Akt-mTOR as well as NF-kB signaling pathways and hence controlling the overexpression of MMP-2. In this report, they mention that fucoidan could be considered as a potential therapeutic agent against the metastasis of invasive human lung cancer cells (Lee, Kim, & Kim, 2012).

X-ray crystal structures of astacin from crayfish, and adamalysins II from snake venom revealed a topological equivalence and virtual identical extended HEXXHXXGXXH zinc-binding segments along with presence of a methionine turn, which is common in zinc peptidases. These features are also found in the MMPs (matrixins) and in the large bacterial *Serratia* proteinase-like peptidases (serralysins) (Bode, Gomis-Rüth, & Stöckler, 1993; Gutiérrez & Rucavado, 2000). In Central and South America, parts of Asia, and Africa, snake bite is a seriously neglected issue (Gutiérrez, Theakston, & Warrell, 2006; Rucavado, Escalante, Shannon, Gutiérrez, & Fox, 2011). The medical practices that are currently existing for the management of post snake bites include antivenoms which are expensive and require a special storage because of the limited shelf life (Fitton, 2011). The highly complex nature of the snake venoms may cause serious medical conditions like swelling, bleeding, blistering, and ultimately the necrosis of skin and muscle. Two major components of the venom zinc-dependent metalloproteinases and myotoxic phospholipases (PLA2) are reported to be responsible for these effects especially the snake bites in South America. The extracellular matrix components are degraded by metalloproteases. The damage of the plasma membrane of skeletal muscle cells is due to myotoxic phospholipases. The metalloproteases and myotoxic phospholipases elicit a large inflammatory reaction which is succeeded by the infiltration of leukocytes resulting in swelling that restricts blood flow, ischemia, and further tissue damage.

Fucoidan from *Ascophyllum nodosum* was found to be effective in inhibiting the PLA2 variants that are found in the venoms of crotalid snakes (Angulo & Lomonte, 2003; Gutiérrez et al., 2006; Mora, Mora, Lomonte, & Gutiérrez, 2008; Rucavado et al., 2011). PLA2 enzyme-mediated lymphatic system damage is reported from the venom of crotalid snakes. In one of the investigations, it was reported that fucoidan could effectively inhibit the PLA2 effects completely and resulted in the complete elimination of myotoxic effects of the venom. However, there was no change in the hemorrhage effects of the venom (Mora et al., 2008). Fucoidan is reported to form complexes with PLA2 enzyme of snake venom and inactivates its toxic effects (Angulo & Lomonte, 2003). The combination of fucoidan with batimastat (a potent MMP inhibitor known to lower the crotalid venom activity) has reduced the presence of intracellular exudates and on the other hand, batimastat reduced the amount of matrix proteins. This combination of fucoidan and batimastat has successfully eliminated most of the pathological responses caused by this venom (Rucavado et al., 2011). With the aid of current day's combinatorial chemistry procedures, best medicinal leads for the control of MMP expression can be formulated.

The expression of MMPs is upregulated by ultraviolet (UV-B) irradiation which in turn induces the cellular signaling transduction pathways that are responsible for the degradation or synthesis inhibition of extracellular matrix in connective tissues (collagenous) resulting in photoaging. The fucoidan derived from brown algae is reported to exhibit strong inhibition ability on MMP-1 expression that is induced by ultraviolet radiations. Human skin fibroblast (HS68) cells were pretreated by various concentrations of fucoidan and then subjected to UV-B irradiation (100 mJ/cm^2). Moreover, in fucoidan-treated cells, the expression of MMP-1 mRNA has been significantly reduced (Moon et al., 2008). As brown edible algae are considered as dietary food stuff, the consumption of brown algae that are rich in fucoidan could reduce the risk of MMP-related diseases.

Molecular weight of the fucoidan has a major impact on its biological effect. A 16-kDa fucoidan fraction obtained from marine algae has exhibited a potential inhibitory effect on the interleukin-1β-mediated gelatinase A secretion and stromelysin-1 stimulation in dermal fibroblasts *in vitro*. These studies have revealed that human leukocyte elastase activity has been decreased by fucoidan fraction, thus rendering protection to the human skin elastic fiber network against serine proteinase effects (Senni et al., 2006). Fucoidans with high molecular weight (HMW) can bind to the FGFs and prevent them from proteolysis (Belford, Hendry, & Parish, 1993). In one of the investigations, there was an enhancement observed in therapeutic revascularization of critical hind limb ischemia in a rat model by LMWF (7 ± 2 kDa) that is obtained by radical depolymerization of HMW (Luyt et al., 2003). In both animal models of cerebral ischemia and human stroke, the MMP-9 plays an important role. Post cerebral ischemia, the matrix degradation is enhanced due to the overexpression of MMP-9 which in turn damages the blood–brain barrier, leading to the increase of the infarct size and relating to hemorrhagic transformation (Dong, Song, Liu, & Guo, 2009). The therapeutic ability of seaweed fucoidans would be a best option in managing the MMP-associated cerebral ischemia. It is also suggested that fucoidans can release the glycosaminoglycan-bound stromal-derived factor-1 (SDF-1) from its tissue storage sites and renders its therapeutic potential. Medullary progenitors are mobilized by SDF-1, resulting in angiogenesis with VEGF and FGF (Salvucci et al., 2002; Sellke et al., 1996).

The pretreatment of immortalized human keratinocyte (HaCaT) cell line with *Costaria costata*-derived fucoidans has shown a significant decrease in the UV-B-induced MMP-1 expression. Moreover, these SPs have significantly reduced the expression of MMP-1 mRNA and inhibited

UV-B–induced MMP-1 promoter activity by 37.3%, 53.3%, and 58.5% at 0.01, 0.1, and 1 μg/ml, respectively, compared to UV-B irradiation alone (Ryu, Qian, Kim, Nam, & Kim, 2009). In another report, it was revealed that *Cladosiphon novaecaledoniae* Kylin (Mozuku)-derived fucoidan extracts have suppressed the expression of MMP-2 and -9 and inhibited cellular invasiveness of human fibrosarcoma HT1080. The suppression effect of fucoidan extracts on invasion and angiogenesis of tumor cells was through the inhibition of expression and secretion of an angiogenesis factor, VEGF (Ye et al., 2005). The pretreated human gastric adenocarcinoma cells (AGS) with fucoidans were observed to have inhibited growth of AGS cells by inducing autophagy, as well as apoptosis. It was also reported that the down-regulation of antiapoptotic Bcl-2 and Bcl-xL expression, loss of mitochondrial membrane potential, activation of caspases, and concomitant degradation of poly-(ADP-ribose) polymerase protein are involved in the fucoidan-induced apoptosis (Park, Kim, Nam, Deuk Kim, & Hyun Choi, 2011). Thus, fucoidans can be of great potential in controlling the expressions of MMPs that regulate the cell proliferation and metastasis and hence can be better dietary supplements in managing cancers.

3. CONCLUSIONS AND FURTHER PROSPECTS

The prominence of the MMPs in the involvement of progression of several pathological conditions that hinders human health has been recognized in recent times. Moreover, the unwanted side effects of the available synthetic MMPIs have made the current day researchers to seek and design inhibitors of MMPs from natural resources. Several biologically active compounds have been reported from terrestrial organisms. However, the structural and chemical diversity of the marine-derived bioactives make the marine resources as the most promising and challenging field of study. Among the marine organisms, there has been a lot of interest in the exploitation of the bioactives occurring in the marine algae. Since ancient times, the marine algae have gained a prominent place in diet especially in the North Pacific Asian region as functional foods. Especially, SPs have been reported with unmatched biological functions. The SPs have shown a promising potential in controlling various kinds of inflammatory responses. As the overexpression of MMPs is one of the outcomes of inflammation, the establishment of new age technologies should be employed in elucidating the mode of action of these bioactives in controlling the MMPs. The knowledge about the interaction of phlorotannins and SPs in controlling the overexpression of MMPs is still at a budding stage. The available molecular biological and advanced scientific technologies gives a current day marine biotechnologist ample scope to elucidate the mechanism of these bioactives in controlling the imbalanced expression of MMPs. Moreover, marine algae harbor many other bioactive such as peptides, minerals, vitamins, and other important compounds that could prove beneficial as MMPIs. Much more scientific investigation is recommended to establish the marine algal metabolites as potential resources for the management of MMPs.

ACKNOWLEDGMENTS

This research was supported by a grant from Marine Bioprocess Research Center of the Marine Biotechnology Program funded by the Ministry of Oceans and Fisheries, Republic of Korea.

Conflict of interest: There is no conflict of interest between the two authors.

REFERENCES

Albuquerque, I., Queiroz, K., Alves, L., Santos, E., Leite, E., & Rocha, H. (2004). Heterofucans from Dictyota menstrualis have anticoagulant activity. *Brazilian Journal of Medical and Biological Research*, *37*(2), 167–171.

Ali, M., Jahangir, M., Saleem, M., Pervez, M., Hameed, S., & Ahmad, V. (2000). Metabolites of marine algae collected from Karachi-coasts of Arabian sea. *Natural Product Sciences*, *6*(2), 61–65.

Angulo, Y., & Lomonte, B. (2003). Inhibitory effect of fucoidan on the activities of crotaline snake venom myotoxic phospholipases A2. *Biochemical Pharmacology*, *66*(10), 1993–2000.

Belford, D. A., Hendry, I. A., & Parish, C. R. (1993). Investigation of the ability of several naturally occurring and synthetic polyanions to bind to and potentiate the biological activity of acidic fibroblast growth factor. *Journal of Cellular Physiology*, *157*(1), 184–189.

Berteau, O., & Mulloy, B. (2003). Sulfated fucans, fresh perspectives: Structures, functions, and biological properties of sulfated fucans and an overview of enzymes active toward this class of polysaccharide. *Glycobiology*, *13*(6), 29R–40R.

Bilan, M. I., & Usov, A. I. (2008). Structural analysis of fucoidans. *Natural Product Communications*, *3*(10), 1639–1648.

Bode, W., Gomis-Rüth, F.-X., & Stöckler, W. (1993). Astacins, serralysins, snake venom and matrix metalloproteinases exhibit identical zinc-binding environments (HEXXHXXGXXH and Met-turn) and topologies and should be grouped into a common family, the 'metzincins'. *FEBS Letters*, *331*(1–2), 134–140.

Bourboulia, D., & Stetler-Stevenson, W. G. (2010). Matrix metalloproteinases (MMPs) and tissue inhibitors of metalloproteinases (TIMPs): Positive and negative regulators in tumor cell adhesion. *Seminars in Cancer Biology, 20*, 161–168, Elsevier.

Brooks, P. C., Strömblad, S., Sanders, L. C., von Schalscha, T. L., Aimes, R. T., Stetler-Stevenson, W. G., et al. (1996). Localization of matrix metalloproteinase MMP-2 to the surface of invasive cells by interaction with integrin $\alpha v \beta 3$. *Cell*, *85*(5), 683–693.

Chen, H., Yan, X., Lin, J., Wang, F., & Xu, W. (2007). Depolymerized products of λ-carrageenan as a potent angiogenesis inhibitor. *Journal of Agricultural and Food Chemistry*, *55*(17), 6910–6917.

Chiu, Y.-H., Chan, Y.-L., Tsai, L.-W., Li, T.-L., & Wu, C.-J. (2012). Prevention of human enterovirus 71 infection by kappa carrageenan. *Antiviral Research*, *95*(2), 128–134.

Chizhov, A. O., Dell, A., Morris, H. R., Haslam, S. M., McDowell, R. A., Shashkov, A. S., et al. (1999). A study of fucoidan from the brown seaweed *Chorda filum*. *Carbohydrate Research*, *320*(1–2), 108–119.

Costa, L. S., Fidelis, G. P., Cordeiro, S. L., Oliveira, R. M., Sabry, D. A., Câmara, R. B. G., et al. (2010). Biological activities of sulfated polysaccharides from tropical seaweeds. *Biomedicine & Pharmacotherapy*, *64*(1), 21–28.

Coussens, L. M., Fingleton, B., & Matrisian, L. M. (2002). Matrix metalloproteinase inhibitors and cancer—Trials and tribulations. *Science*, *295*(5564), 2387–2392.

Cumashi, A., Ushakova, N. A., Preobrazhenskaya, M. E., D'Incecco, A., Piccoli, A., Totani, L., et al. (2007). A comparative study of the anti-inflammatory, anticoagulant, antiangiogenic, and antiadhesive activities of nine different fucoidans from brown seaweeds. *Glycobiology*, *17*(5), 541–552.

Damonte, E. B., Matulewicz, M. C., & Cerezo, A. S. (2004). Sulfated seaweed polysaccharides as antiviral agents. *Current Medicinal Chemistry*, *11*(18), 2399–2419.

de Azevedo, C. G., Bezerra, M. E. B., Santos Mda, G., Souza, L. A., Marques, C. T., Benevides, N. M. B., et al. (2009). Heparinoids algal and their anticoagulant, hemorrhagic activities and platelet aggregation. *Biomedicine & Pharmacotherapy*, *63*(7), 477–483.

Dong, X., Song, Y. N., Liu, W. G., & Guo, X. L. (2009). MMP-9, a potential target for cerebral ischemic treatment. *Current Neuropharmacology*, *7*(4), 269–275.

Egeblad, M., & Werb, Z. (2002). New functions for the matrix metalloproteinases in cancer progression. *Nature Reviews. Cancer*, *2*(3), 161–174.

Farias, W. R., Nazareth, R. A., & Mourão, P. A. (2001). Dual effects of sulfated D-galactans from the red algae Botryocladia occidentalis preventing thrombosis and inducing platelet aggregation. *Thrombosis and Haemostasis*, *86*(6), 1540–1546.

Farias, E. H., Pomin, V. H., Valente, A.-P., Nader, H. B., Rocha, H. A., & Mourão, P. A. (2008). A preponderantly 4-sulfated, 3-linked galactan from the green alga Codium isthmocladum. *Glycobiology*, *18*(3), 250–259.

Fitton, J. H. (2011). Therapies from fucoidan; multifunctional marine polymers. *Marine Drugs*, *9*(10), 1731–1760.

Ghosh, T., Chattopadhyay, K., Marschall, M., Karmakar, P., Mandal, P., & Ray, B. (2009). Focus on antivirally active sulfated polysaccharides: From structure–activity analysis to clinical evaluation. *Glycobiology*, *19*(1), 2–15.

Gill, S. E., & Parks, W. C. (2008). Metalloproteinases and their inhibitors: Regulators of wound healing. *The International Journal of Biochemistry & Cell Biology*, *40*(6–7), 1334–1347.

Girond, S., Crance, J., Van Cuyck-Gandre, H., Renaudet, J., & Deloince, R. (1991). Antiviral activity of carrageenan on hepatitis A virus replication in cell culture. *Research in Virology*, *142*(4), 261–270.

Gómez-Ordóñez, E., Jiménez-Escrig, A., & Rupérez, P. (2010). Dietary fibre and physicochemical properties of several edible seaweeds from the northwestern Spanish coast. *Food Research International*, *43*(9), 2289–2294.

Gross, J., & Lapiere, C. M. (1962). Collagenolytic activity in amphibian tissues: A tissue culture assay. *Proceedings of the National Academy of Sciences of the United States of America*, *48*(6), 1014.

Gupta, S. C., Kim, J. H., Prasad, S., & Aggarwal, B. B. (2010). Regulation of survival, proliferation, invasion, angiogenesis, and metastasis of tumor cells through modulation of inflammatory pathways by nutraceuticals. *Cancer and Metastasis Reviews*, *29*(3), 405–434.

Gutiérrez, J. M., & Rucavado, A. (2000). Snake venom metalloproteinases: Their role in the pathogenesis of local tissue damage. *Biochimie*, *82*(9–10), 841–850.

Gutiérrez, J. M., Theakston, R. D. G., & Warrell, D. A. (2006). Confronting the neglected problem of snake bite envenoming: The need for a global partnership. *PLoS Medicine*, *3*(6), e150.

Ha, K. T., Kim, J. K., Kang, S. K., Kim, D. W., Lee, Y. C., Kim, H. M., et al. (2004). Inhibitory effect of Sihoga-Yonggol-Moryo-Tang on matrix metalloproteinase-2 and -9 activities and invasiveness potential of hepatocellular carcinoma. *Pharmacological Research*, *50*(3), 279–285.

Hayakawa, Y., Hayashi, T., Lee, J.-B., Srisomporn, P., Maeda, M., Ozawa, T., et al. (2000). Inhibition of thrombin by sulfated polysaccharides isolated from green algae. *Biochimica et Biophysica Acta*, *1543*(1), 86–94.

Hlawaty, H., Suffee, N., Sutton, A., Oudar, O., Haddad, O., Ollivier, V., et al. (2011). Low molecular weight fucoidan prevents intimal hyperplasia in rat injured thoracic aorta through the modulation of matrix metalloproteinase-2 expression. *Biochemical Pharmacology*, *81*(2), 233–243.

Hodgson, J. (1995). Remodeling MMPIs. *Biotechnology*, *13*(6), 554–557.

Hoffman, R., Burns, W. W., & Paper, D. H. (1995). Selective inhibition of cell proliferation and DNA synthesis by the polysulphated carbohydrate ι-carrageenan. *Cancer Chemotherapy and Pharmacology*, *36*(4), 325–334.

Holdt, S. L., & Kraan, S. (2011). Bioactive compounds in seaweed: Functional food applications and legislation. *Journal of Applied Phycology*, *23*(3), 543–597.

Kessenbrock, K., Plaks, V., & Werb, Z. (2010). Matrix metalloproteinases: Regulators of the tumor microenvironment. *Cell*, *141*(1), 52–67.

Knutsen, S., Myslabodski, D., Larsen, B., & Usov, A. (1994). A modified system of nomenclature for red algal galactans. *Botanica Marina*, *37*(2), 163–170.

Lambert, E., Dassé, E., Haye, B., & Petitfrère, E. (2004). TIMPs as multifacial proteins. *Critical Reviews in Oncology/Hematology*, *49*(3), 187–198.

Lee, H., Kim, J.-S., & Kim, E. (2012). Fucoidan from seaweed Fucus vesiculosus inhibits migration and invasion of human lung cancer cell via PI3K-Akt-mTOR pathways. *PLoS One*, *7*(11), e50624.

Lee, M. H., & Murphy, G. (2004). Matrix metalloproteinases at a glance. *Journal of Cell Science*, *117*(18), 4015–4016.

Li, B., Lu, F., Wei, X., & Zhao, R. (2008). Fucoidan: Structure and bioactivity. *Molecules*, *13*(8), 1671–1695.

Liotta, L., Tryggvason, K., Garbisa, S., Hart, I., Foltz, C., & Shafie, S. (1980). Metastatic potential correlates with enzymatic degradation of basement membrane collagen. *Nature*, *284*(5751), 67–68.

Luyt, C. E., Meddahi-Pellé, A., Ho-Tin-Noe, B., Colliec-Jouault, S., Guezennec, J., Louedec, L., et al. (2003). Low-molecular-weight fucoidan promotes therapeutic revascularization in a rat model of critical hindlimb ischemia. *The Journal of Pharmacology and Experimental Therapeutics*, *305*(1), 24.

Malemud, C. J. (2006). Matrix metalloproteinases (MMPs) in health and disease: An overview. *Frontiers in Bioscience*, *11*(1), 1696–1701.

Matsubara, K., Matsuura, Y., Bacic, A., Liao, M., Hori, K., & Miyazawa, K. (2001). Anticoagulant properties of a sulfated galactan preparation from a marine green alga, Codium cylindricum. *International Journal of Biological Macromolecules*, *28*(5), 395.

Mayer, A., Rodríguez, A. D., Berlinck, R. G., & Fusetani, N. (2011). Marine pharmacology in 2007–8: Marine compounds with antibacterial, anticoagulant, antifungal, antiinflammatory, antimalarial, antiprotozoal, antituberculosis, and antiviral activities; affecting the immune and nervous system, and other miscellaneous mechanisms of action. *Comparative Biochemistry and Physiology, Part C: Toxicology & Pharmacology*, *153*(2), 191–222.

McLean, M. W., & Williamson, F. B. (1979). Kappa-Carrageenase from Pseudomonas carrageenovora. *European Journal of Biochemistry*, *93*(3), 553–558.

Moon, H. J., Lee, S. R., Shim, S. N., Jeong, S. H., Stonik, V. A., Rasskazov, V. A., et al. (2008). Fucoidan inhibits UVB-induced MMP-1 expression in human skin fibroblasts. *Biological & Pharmaceutical Bulletin*, *31*(2), 284–289.

Mora, J., Mora, R., Lomonte, B., & Gutiérrez, J. M. (2008). Effects of Bothrops asper snake venom on lymphatic vessels: Insights into a hidden aspect of envenomation. *PLoS Neglected Tropical Diseases, 2*(10), e318.

Mourão, P. (2007). A carbohydrate-based mechanism of species recognition in sea urchin fertilization. *Brazilian Journal of Medical and Biological Research, 40*(1), 5–17.

Mulloy, B. (2005). The specificity of interactions between proteins and sulfated polysaccharides. *Anais da Academia Brasileira de Ciências, 77*(4), 651–664.

Olson, M. W., Gervasi, D. C., Mobashery, S., & Fridman, R. (1997). Kinetic analysis of the binding of human matrix metalloproteinase-2 and -9 to tissue inhibitor of metalloproteinase (TIMP)-1 and TIMP-2. *Journal of Biological Chemistry, 272*(47), 29975–29983.

Opoku, G., Qiu, X., & Doctor, V. (2006). Effect of oversulfation on the chemical and biological properties of kappa carrageenan. *Carbohydrate Polymers, 65*(2), 134–138.

Overall, C. M., & López-Otín, C. (2002). Strategies for MMP inhibition in cancer: Innovations for the post-trial era. *Nature Reviews. Cancer, 2*(9), 657–672.

Page-McCaw, A., Ewald, A. J., & Werb, Z. (2007). Matrix metalloproteinases and the regulation of tissue remodelling. *Nature Reviews. Molecular Cell Biology, 8*(3), 221–233.

Park, H. S., Kim, G. Y., Nam, T. J., Deuk Kim, N., & Hyun Choi, Y. (2011). Antiproliferative activity of fucoidan was associated with the induction of apoptosis and autophagy in AGS human gastric cancer cells. *Journal of Food Science, 76*(3), T77–T83.

Parks, W. C., Wilson, C. L., & López-Boado, Y. S. (2004). Matrix metalloproteinases as modulators of inflammation and innate immunity. *Nature Reviews. Immunology, 4*(8), 617–629.

Patankar, M. S., Oehninger, S., Barnett, T., Williams, R. L., & Clark, G. (1993). A revised structure for fucoidan may explain some of its biological activities. *The Journal of Biological Chemistry, 268*(29), 21770.

Pomin, V. H., & Mourão, P. A. (2008). Structure, biology, evolution, and medical importance of sulfated fucans and galactans. *Glycobiology, 18*(12), 1016–1027.

Rucavado, A., Escalante, T., Shannon, J., Gutiérrez, J. M., & Fox, J. W. (2011). Proteomics of wound exudate in snake venom–induced pathology: Search for biomarkers to assess tissue damage and therapeutic success. *Journal of Proteome Research, 10*(4), 1987–2005.

Rundhaug, J. E. (2005). Matrix metalloproteinases and angiogenesis. *Journal of Cellular and Molecular Medicine, 9*(2), 267–285.

Ryu, B., Qian, Z.-J., Kim, M.-M., Nam, K. W., & Kim, S.-K. (2009). Anti-photoaging activity and inhibition of matrix metalloproteinase (MMP) by marine red alga, Corallina pilulifera methanol extract. *Radiation Physics and Chemistry, 78*(2), 98–105.

Salvucci, O., Yao, L., Villalba, S., Sajewicz, A., Pittaluga, S., & Tosato, G. (2002). Regulation of endothelial cell branching morphogenesis by endogenous chemokine stromal-derived factor-1. *Blood, 99*(8), 2703.

Santos, G. (1989). Carrageenans of species of *Eucheuma* J. Agardh and *Kappaphycus* Doty (Solieriaceae, Rhodophyta). *Aquatic Botany, 36*(1), 55–67.

Sellke, F. W., Li, J., Stamler, A., Lopez, J. J., Thomas, K. A., & Simons, M. (1996). Angiogenesis induced by acidic fibroblast growth factor as an alternative method of revascularization for chronic myocardial ischemia. *Surgery, 120*(2), 182–188.

Senni, K., Gueniche, F., Foucault-Bertaud, A., Igondjo-Tchen, S., Fioretti, F., Colliec-Jouault, S., et al. (2006). Fucoidan a sulfated polysaccharide from brown algae is a potent modulator of connective tissue proteolysis. *Archives of Biochemistry and Biophysics, 445*(1), 56–64.

Seo, U. K., Lee, Y. J., Kim, J. K., Cha, B. Y., Kim, D. W., Nam, K. S., et al. (2005). Large-scale and effective screening of Korean medicinal plants for inhibitory activity on matrix metalloproteinase-9. *Journal of Ethnopharmacology, 97*(1), 101–106.

Shuman Moss, L. A., Jensen-Taubman, S., & Stetler-Stevenson, W. G. (2012). Matrix metalloproteinases: Changing roles in tumor progression and metastasis. *The American Journal of Pathology, 181*(6), 1895–1899.

Silva, T., Alves, L., de Queiroz, K., Santos, M., Marques, C., Chavante, S., et al. (2005). Partial characterization and anticoagulant activity of a heterofucan from the brown seaweed Padina gymnospora. *Brazilian Journal of Medical and Biological Research, 38*(4), 523–533.

Smestad Paulsen, B. (2002). Biologically active polysaccharides as possible lead compounds. *Phytochemistry Reviews, 1*(3), 379–387.

Talarico, L., Pujol, C., Zibetti, R., Faría, P., Noseda, M., Duarte, M., et al. (2005). The antiviral activity of sulfated polysaccharides against dengue virus is dependent on virus serotype and host cell. *Antiviral Research, 66*(2–3), 103–110.

Thomas, N. V., & Kim, S. K. (2010). Metalloproteinase inhibitors: Status and scope from marine organisms. *Biochemistry Research International, 2010,* 845975.

Thomas, N. V., & Kim, S. K. (2013). Beneficial effects of marine algal compounds in cosmeceuticals. *Marine Drugs, 11*(1), 146–164.

Usov, A. I., & Bilan, M. I. (2009). Fucoidans—Sulfated polysaccharides of brown algae. *Russian Chemical Reviews, 78*(8), 785.

Velinov, N., Poptodorov, G., Gabrovski, N., & Gabrovski, S. (2010). The role of matrix metalloproteinases in the tumor growth and metastasis. *Khirurgiia, 2010*(1), 44–49.

Wang, W., Zhang, P., Yu, G.-L., Li, C.-X., Hao, C., Qi, X., et al. (2012). Preparation and anti-influenza A virus activity of k-carrageenan oligosaccharide and its sulphated derivatives. *Food Chemistry, 133*(2012), 880–888.

Yang, L., & Zhang, L.-M. (2009). Chemical structural and chain conformational characterization of some bioactive polysaccharides isolated from natural sources. *Carbohydrate Polymers, 76*(3), 349–361.

Ye, J., Li, Y., Teruya, K., Katakura, Y., Ichikawa, A., Eto, H., et al. (2005). Enzyme-digested fucoidan extracts derived from seaweed Mozuku of Cladosiphon novae-caledoniaekylin inhibit invasion and angiogenesis of tumor cells. *Cytotechnology, 47*(1), 117–126.

Yu, Q., & Stamenkovic, I. (2000). Cell surface-localized matrix metalloproteinase-9 proteolytically activates TGF-β and promotes tumor invasion and angiogenesis. *Genes & Development, 14*(2), 163–176.

Yu, Q., Yan, J., Wang, S., Ji, L., Ding, K., Vella, C., et al. (2012). Antiangiogenic effects of GFP08, an agaran-type polysaccharide isolated from Grateloupia filicina. *Glycobiology, 22*(10), 1343–1352.

Yuan, H., Zhang, W., Li, X., Lü, X., Li, N., Gao, X., et al. (2005). Preparation and in vitro antioxidant activity of κ-carrageenan oligosaccharides and their oversulfated, acetylated, and phosphorylated derivatives. *Carbohydrate Research, 340*(4), 685–692.

Zhang, C., & Kim, S.-K. (2009). Matrix metalloproteinase inhibitors (MMPIs) from marine natural products: The current situation and future prospects. *Marine Drugs, 7*(2), 71–84.

CHAPTER ELEVEN

Anticancer Effects of Fucoidan

Kalimuthu Senthilkumar*,†, Se-Kwon Kim‡,1

*Specialized Graduate School Science and Technology Convergence, Department of Marine Bio Convergence Science, Pukyong National University, Busan, South Korea
†Department of Nuclear Medicine, Kyungpook National University School of Medicine, Daegu, South Korea
‡Department of Marine-bio Convergence Science, Specialized Graduate School Science and Technology Convergence, Marine Bioprocess Research Center, Pukyong National University, Busan, South Korea
1Corresponding author: e-mail address: sknkim@pknu.ac.kr

Contents

1. Introduction 195
2. Seaweed Polysaccharides 196
   2.1 Fucoidan 197
   2.2 Fucoidan structure and function 198
3. Fucoidan and Cancer 200
   3.1 Anticancer effect 200
   3.2 Role of fucoidan on metastasis, angiogenesis, and signaling mechanism 204
4. Conclusions 206
Acknowledgment 207
References 207

Abstract

Recently, there has been an increased interest in the pharmacologically active natural compounds isolated and used for remedies of various kinds of diseases, including cancer. The great deal of interest has been developed to isolate bioactive compounds from marine resources because of their numerous health beneficial effects. Among marine resources, marine algae are valuable sources of structurally diverse bioactive compounds. Fucoidan is a sulfated polysaccharide derived from brown seaweeds and has been used as an ingredient in some dietary supplement products. Fucoidan has various biological activities including antibacterial, antioxidant, anti-inflammatory, anticoagulant, and anti-tumor activities. So this chapter deals with anticancer effects of fucoidan.

1. INTRODUCTION

Cancer is a leading cause of death worldwide. It is a diverse group of diseases characterized by the uncontrolled proliferation of anaplastic cells which tend to invade surrounding tissues and metastasize to other tissues and organs. Cancer results from a mutation in the chromosomal DNA of a normal cell, which can be triggered by both external factors (tobacco, alcohol, chemicals, infectious agents, and radiation) and internal factors (hormones, immune conditions, inherited mutations, and mutations occurring in metabolism), promoting cancer progression. Taking into account the rising trend of the incidence of cancers of various organs, effective therapies are urgently needed to control human malignancies. In the past few decades, various drugs arise from marine environment.

Life began in the sea covering over 70% of the Earth's surface. The marine environment contains a diverse number of plants, animals, and microorganisms, which have a wide diversity of natural products (Costantino, Fattorusso, Menna, & Taglialatela-Scafati, 2004). Among these natural sources, marine algae have great frontier for pharmaceutical and medical research. Recent studies in the field of cancer research have revealed promising compounds, isolated from natural sources, with proven anticancer activity. The phaeophyceae or brown algae are a large group of mostly marine multicellular algae, including many seaweeds. They play an important role in marine environments, both as food and for the habitats they form. Nowadays, the field of marine natural products becomes more sophisticated. Seaweeds produce a variety of biologically active components with different structures and interesting functional properties. Seaweeds have great potential as a supplement in functional food or for the extraction of compounds. Seaweeds are known for their richness in polysaccharides, minerals, and certain vitamins, but they also contain bioactive substances including polysaccharides, proteins, lipids, and polyphenols (Chandini, Ganesan, Suresh, & Bhaskar, 2008; Holdt & Kraan, 2011) and used for the development of new pharmaceutical agents (Bhadury, Mohammad, & Wright, 2006).

2. SEAWEED POLYSACCHARIDES

Brown seaweeds belong to a very large group and it is the second most abundant group of seaweeds (Mestechkina & Shcherbukhin, 2010). Polysaccharides produced by seaweeds expand the economically important and global industries. Significant amounts of seaweed-derived polysaccharides are used in food, pharmaceuticals, and other products for human consumption (Renn, 1997). Over the past decade, bioactive sulfated polysaccharides isolated from brown seaweeds have attracted much attention in the fields of pharmacology and biochemistry. Functional polysaccharides such as fucans and alginic acid derivatives produced by brown seaweeds are known to exhibit different biological properties including anticoagulant, antiinflammatory, antiviral, and antitumoral activities (Boisson-Vidal et al., 1995; Costa et al., 2010; Lee, Athukorala, Lee, & Jeon, 2008). Sulfated polysaccharides, fucoidans have been isolated from different brown algal species such as *Ecklonia cava*, *Ascophyllum nodosum*, and *Undaria pinnatifida* (Athukorala, Jung, Vasanthan, & Jeon, 2006; Matou, Helley, Chabut, Bros, & Fischer, 2002). Seaweeds have low lipid, high carbohydrate, and more dietary fibers. The cell wall polysaccharides mainly consist of cellulose and hemicelluloses, neutral polysaccharides, and they are thought to physically support the thallus in water. Green algae contain sulfuric acid polysaccharides, sulfated galactans, and xylans; brown algae contain alginic acid, fucoidan (sulfated fucose), laminarin (β-1,3 glucan), and sargassan; and red algae contain agars, carrageenans, xylans, floridean starch (amylopectin-like glucan), water-soluble sulfated galactan, as well as porphyran as mucopolysaccharides located in the intercellular spaces (Chandini et al., 2008; Murata & Nakazoe, 2001). The components of galactose, glucose, mannose, fructose, xylose, fucose, and arabinose were found in the total sugars in the hydrolysates. The glucose content was 65%, 30%, and 20% of the total sugars in an autumn sample of 50 individual plants of Saccharina, Fucus (serratus and spiralis), and Ascophyllum, respectively (Jensen, 1956). Several other polysaccharides are present in and utilized from seaweed, for example, furcellaran, funoran, ascophyllan, and sargassan. Brown algae mainly contain L-fucose and sulfate ester groups containing polysaccharides called fucoidan.

2.1. Fucoidan

During the past decade, numerous bioactive polysaccharides with interesting functional properties have been discovered from marine algae (Patel, 2012). Fucoidan, whose molecular weight average is about 20,000, is a sulfated polysaccharide extracted from brown marine algae and consists of L-fucose and sulfate ester groups. It is found mainly in various species of brown algae and brown seaweed such as mozuku, kombu, limu moui, bladderwrack, wakame, and hijiki. Fucoidan is a term used for a class of sulfated, fucose-rich, polysaccharides found in the fibrillar cell walls and intercellular spaces of brown seaweeds. Structurally, fucoidan is a heparin-like molecule with a substantial percentage of L-fucose, sulfated ester groups, as well as small proportions of D-xylose, D-galactose, D-mannose, and glucuronic acid (Gideon & Rengasamy, 2008; Jiao, Yu, Zhang, & Ewart, 2011). The polysaccharide was named as "fucoidin" when it was first isolated from marine brown algae by Kylin in 1913 (Kylin, 1913). Now it is named as "fucoidan" according to IUPAC rules, but also called as fucan, fucosan, or sulfated fucan (Berteau & Mulloy, 2003). Most of the fucoidans are complex in their chemical composition.

2.2. Fucoidan structure and function

There is inconsistency among the sulfated monosaccharides (i.e., glucose (Glc), galactose (Gal), the corresponding N-acetyl (NAc) amines GlcNAc, GalNAc, and mannose among others), the total number of sulfate moieties, and the hydroxyl(s) to which the sulfate group(s) are linked (i.e., 2-O, 3-O, 4-O, and 6-O sulfates), furthermore, the varied structure of the underlying oligosaccharide moiety. The most studied FCSPs, originally called fucoidin, fucoidan, or just fucans, have a backbone built of $(1 \rightarrow 3)$-linked α-L-fucopyranosyl residues or of alternating $(1 \rightarrow 3)$- and $(1 \rightarrow 4)$-linked α-L-fucopyranosyl residues (Patankar, Oehninger, Barnett, Williams, & Clark, 1993; Percival & McDowell, 1967). These fucopyranosyl residues may be substituted with short-fucoside side chains or sulfate groups at C-2 or C-4, and may also carry other minor substitutions, for example, acetate, xylose, mannose, glucuronic acid, galactose, or glucose (Bilan & Usov, 2009; Duarte, Cardoso, Noseda, & Cerezo, 2001; Tako, Yoza, & Tohma, 2000). FCSPs also include sulfated galactofucans with backbones built of $(1 \rightarrow 6)$-β-D-galacto- and/or $(1 \rightarrow 2)$-β-D-mannopyranosyl units. In addition to sulfate, these backbone residues may be substituted with fucosides, single fucose substitutions, and/or glucuronic acid, xylose or glucose substitutions (Duarte et al., 2001). Fucoidan is extracted from brown seaweed algae in which the percentage of L-fucose ranged from 12.6% to 36.0%, and the percentage of sulfate content from 8% to 25%. The structure of fucoidan is depicted in Fig. 11.1. Recent studies suggested that low molecular weight fucoidan (LMWF) has more biological actions than native fucoidan. LMWF, a sulfated polysaccharide, varies with its molecular weight, which is generally classified as low (<10 kDa), medium (10–10,000 kDa), or high >10,000 kDa (Matsubara et al., 2005).

Fucoidan or FCSPs from seaweed have many potential biological functions including antitumor and immunomodulatory (Ale, Maruyama, Tamauchi, Mikkelsen, & Meyer, 2011; Alekseyenko et al., 2007; Maruyama, Tamauchi, Iizuka, & Nakano, 2006), antivirus (Makarenkova, Deriabin, L'vov, Zviagintseva, & Besednova, 2010), antithrombotic and anticoagulant (Zhu et al., 2010), anti-inflammatory (Semenov et al., 1998),

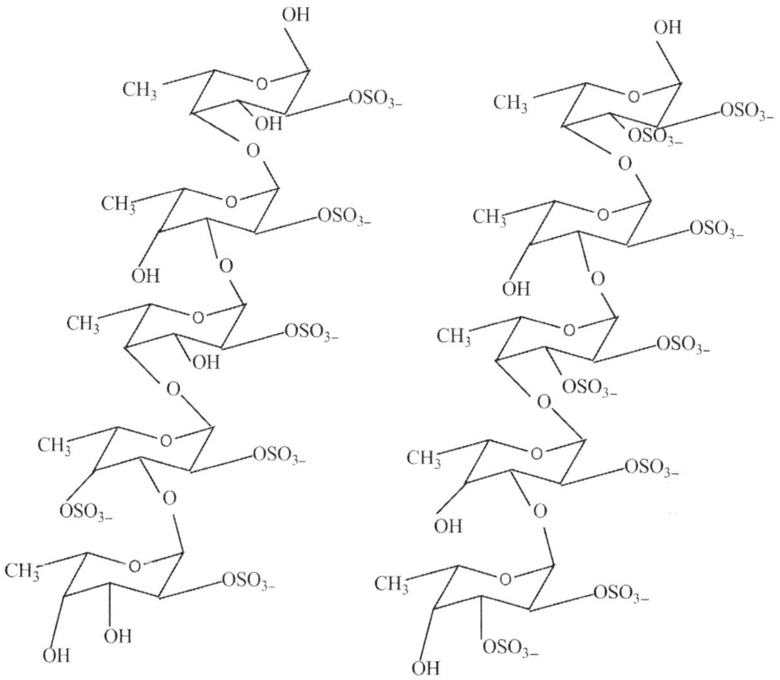

Figure 11.1 Structure of fucoidans.

and antioxidant effects (Wang, Zhang, Zhang, Song, & Li, 2010), as well as their effects against renal (Veena, Josephine, Preetha, Varalakshmi, & Sundarapandiyan, 2006), hepatic (Hayakawa & Nagamine, 2009), uropathic disorders (Zhang et al., 2005), anticoagulant, antiviral, and anticancer properties (Chevolot, Mulloy, Ratiskol, Foucault, & Colliec-Jouault, 2001; Zhuang, Itoh, Mizuno, & Ito, 1995). Fucoidan is an excellent natural antioxidant and has significant antioxidant activity. It also stimulates the immune system and antiviral, antiherpetic activity, and regulates cell signaling molecules in various types of cancer (Senthilkumar, Manivasagan, Venkatesana, & Kim, 2013). LMWF has anti-inflammatory properties, and anticomplementary activities with both inhibition of leukocyte accumulation and connective tissue proteolysis and used for treating some inflammatory diseases (Senni et al., 2006). LMWF also promote tissue-rebuilding parameters such as signaling by heparin-binding growth factors (FGF-2, VEGF) and collagen processing in fibroblasts, smooth muscle cells, or endothelial cells in culture. Also LMWF mimics and restores the properties of bone noncollagenous matrix (proteoglycans, glycoproteins), eliminated by drastic purification process during design of the biomaterial, to regulate soluble factors bioavailability (Changotade et al., 2008).

3. FUCOIDAN AND CANCER
3.1. Anticancer effect

Cancer development, a dynamic and long-term process, involves many complex factors with a stepwise progression that ultimately leads to metastasis. The three critical steps in this process for several types of human cancer formation are initiation, promotion, and progression. The National Cancer Institute estimates that approximately 11.4 million Americans with a history of cancer were alive in January 2006. In 2012, about 577,190 Americans are expected to die of cancer, more than 1500 people a day. Cancer is the second most common cause of death in the United States, exceeded only by heart disease, accounting for nearly 1 of every 4 deaths (Source: Cancer Facts and Figs. 2012 of the American Cancer Society). Accordingly, research must continue to progress to improve existing therapies and to develop novel cures. The therapeutic strategies such as chemotherapy, radiation therapy, surgery, or combinations have been used to treat cancer. Unfortunately, these treatments produce various side effects (Grossi et al., 2010; Schneider, Stipper, & Besserer, 2010). Epidemiological studies have provided convincing evidence that natural compounds can modify this process. The therapeutic potential of natural bioactive compounds such as polysaccharides, especially fucoidan is now well documented, and this activity combined with natural biodiversity will allow the development of a new generation of therapeutic measures against cancer (Jiao et al., 2011).

Fucoidan inhibits proliferation and induces apoptosis in human lymphoma HS-Sultan cell lines (Aisa et al., 2005). Fucoidans from brown seaweeds *E. cava*, *Sargassum hornery*, and *Costaria costata* showed anticancer effect on human melanoma and colon cancer cells (Ermakova et al., 2011). Human malignant melanoma cancer cell (SK-MEL-28 and SK-MEL-5) growth was inhibited by fucoidan, isolated from *Fucus evanescens* (Anastyuk, Imbs, Shevchenko, Dmitrenok, & Zvyagintseva, 2012). Fucoidans from *L. saccharina*, *L. digitata*, *Fucus serratus*, *F. distichus*, and *F. vesiculosus* strongly blocked MDA-MB-231 breast carcinoma cell adhesion and implications in tumor metastasis (Cumashi et al., 2007). Fucoidan after single and repeated administration in a dose of 10 mg/kg produced moderate antitumor and antimetastatic effects (Alekseyenko et al., 2007). The mice were fed with the diet containing fucoidan for 40 days. Mekabu fucoidan inhibited tumors by 65.4% (Maruyama et al., 2006). Native and oversulfated FCSPs derived from *Cladosiphon okamuranus* (Chordariales) possess antiproliferative activity in human leukemia cell line (U937) (Teruya, Konishi, Uechi, Tamaki, & Tako, 2007). Sulfated polysaccharides from brown seaweeds *S. japonica* and *U. pinnatifida* possessed high antitumor activity and inhibit proliferation and colony formation of breast cancer and melanoma cancer cell lines (Vishchuk, Ermakova, & Zvyagintseva, 2011). LMWF inhibits human carcinoma cells, including HeLa cervix adenocarcinoma, HT1080 fibrosarcoma, K562 leukaemia, U937 lymphoma, A549 lung adenocarcinoma, and HL-60 (Zhang, Teruya, Eto, & Shirahata, 2011). LMWF inhibits invasion and angiogenesis of HT 1080 fibrosarcoma cells and induces apoptosis in MCF-7 cancer cells (Ye et al., 2005; Zhang et al., 2011). LMWF induces apoptosis through mitochondrial-mediated pathways in MDA-MB-231 breast cancer cells and also evidenced the interrelated roles of Ca^{2+} homeostasis, mitochondrial dysfunction, and caspase activation (Zhang, Teruya, Eto, & Shirahata, 2013). LMWF with sulfate content higher than 20% was found to exert profound anticoagulant activity as well as antiproliferative effects on fibroblast cell line (CCL39) in a dose-dependent manner (Haroun-Bouhedja, Ellouali, Sinquin, & Boisson-Vidal, 2000).

The cell cycle is a high-energy demanding process involving four sequential phases that go from quiescence (G0 phase) to proliferation (G1, S, G2, and M phases) and back to quiescence (Norbury & Nurse, 1992). Fucoidan increased the G0/G1-phase population in hepatocarcinoma cell line (Huh7) accompanied by a decrease in the S phase, suggesting that fucoidan may cause the cell cycle arrest at the G0/G1 phase (Nagamine et al., 2009). Fucoidan suppressed cell proliferation and arrest cell cycle in HCC cell lines (HAK-1A, KYN-2, KYN-3), revealing an increased number of cells in the G2/M phase at 72 h after the addition of the fucoidan (22.5 μg/ml) (Fukahori et al., 2008). Fucoidan induced the accumulation of cells in G1/S phase of the cell cycle on HUT-102 cells (T-cell lymphoma) (Haneji et al., 2005) and nonsmall cell human bronchopulmonary carcinoma (NSCLC-N6) cells (Riou et al., 1996). The growth inhibitory function of fucoidan on human T-cell leukemia virus type 1 (HTLV-1)-infected T cells and MCF-7 cells was mainly due to the induction of cell cycle arrest and apoptosis (Banafa & Roshan, 2013; Haneji et al., 2005). Fucoidan suppressed cell proliferation of HCC, cholangiocarcinoma, and gallbladder carcinoma cell lines in dose- and time-dependent manner by induction of apoptosis and the inhibition of cell cycle (Fukahori et al., 2008).

Apoptosis is critically important for the survival of multicellular organisms (Lockshin & Zakeri, 2007). The process of programmed cell death, or apoptosis, is generally characterized by distinct morphological characteristics and energy-dependent biochemical mechanisms (Johnstone, Ruefli, & Lowe, 2002). Apoptosis is mediated through two major pathways: the extrinsic (death receptor) and intrinsic (mitochondrial) pathways represent the two major well-studied apoptotic processes (Duprez, Wirawan, Berghe, & Vandenabeele, 2009; Sprick & Walczak, 2004). The extrinsic or death receptor pathway, stimulation of death receptors, such as Fas and tumor necrosis factor receptor-1, leads to clustering and formation of a death-inducing signaling complex, which includes the adaptor protein Fas-associated death domain and initiator caspases, such as caspase-8. Activated caspase-8 directly activates downstream effector caspases, such as caspases-3 and -7 (Ashkenazi & Dixit, 1998). Caspase-8 can cleave Bid (Bcl-2 interacting protein) into tBid (truncated Bid), and interacts with proapoptotic protein Bax, and the accumulation of Bax in mitochondria promotes cytochrome c released into the cytosol (Desagher et al., 1999; Eskes, Desagher, Antonsson, & Martinou, 2000; Li & Yuan, 2008). The intrinsic or mitochondrial pathway, the release of several mitochondrial intermembrane space proteins, such as cytochrome c, which associate with Apaf-1 and procaspase-9 to form the apoptosome. Activated caspase-9 can cleave and activate effector caspases, such as caspases-3 and -7 (Hengartner, 2000). Studies showed that fucoidan, extracted from *C. okamuranus*, has strongly antiproliferative and apoptotic effects on MCF-7 cells in a dose-dependent manner. However, fucoidan did not affect proliferation of normal cells of human mammalian epithelial cells. The characteristics of apoptotic cell death are induction of chromatin condensation, fragmentation of nuclei and DNA, and cleavage of specific proteins. Fucoidan induced accumulation of sub-G1 population, chromatin condensation, and internucleosomal fragmentation of DNA (Yamasaki-Miyamoto, Yamasaki, Tachibana, & Yamada, 2009). Activation of caspase-7 and PARP cleavage are hallmarks of apoptosis in MCF-7 cells (Kaufmann, Desnoyers, Ottaviano, Davidson, & Poirier, 1993). Cleavage of PARP and activation of caspase-7 were induced after treatment with fucoidan in MCF-7. Caspase-3 is known to be activated and plays a pivotal role in fucoidan-induced cell death (Aisa et al., 2005; Teruya et al., 2007). Caspase-7 is required by MCF-7 cells and activation of caspase-3 is not necessarily a prerequisite for fucoidan-induced apoptosis (Yamasaki-Miyamoto et al., 2009). Fucoidan extract increased mitochondrial depolarization by upregulating the expression of proapoptotic proteins Bax, Bad, and down-regulating the expression of antiapoptotic proteins Bcl-2 and Bcl-xl in MCF-7 cells (Zhang et al., 2011).

The FCSPs isolated from *F. vesiculosus* enhance mitochondrial membrane permeability of human colon cancer cells *in vitro*, and induce cytochrome *c* and Smac/Diablo released from the mitochondria (Kim, Park, Lee, & Park, 2010). Studies from Jin, Song, Kim, Park, and Kwak (2010) reported the intracellular levels of apoptotic proteins modulated in fucoidan-treated HL-60 cells. The levels of the active forms of caspases-8, -9, and -3 increased in response to fucoidan in a dose-dependent manner and also the cleavage of PARP, a typical substrate for caspase-3. Apoptosis is induced by activation of the caspase pathway and downregulation of the ERK pathway (Aisa et al., 2005; Haneji et al., 2005). LMWF induces a caspase-independent, mitochondrial-mediated apoptotic pathway in ER-positive MCF-7 cells (Zhang et al., 2011).

LMWF induces a sustained collapse of mitochondrial membrane potency, release of cytochrome *c*, downregulation of antiapoptotic proteins Bcl-2, Mcl-1, and Bcl-xl, and activation of caspases-9, -7, and -3, thereby inducing apoptosis in MDA-MB-231 cells (Zhang et al., 2013). Another way, changes in Ca^{2+} signaling can affect cell proliferation and modulate apoptosis. LMWF increases in the intracellular Ca^{2+} level in MDA-MB-231 cells, suggesting that the disruption of Ca^{2+} homeostasis played an important role in the apoptotic process (Zhang et al., 2013). Calpain is a protein belonging to the family of calcium-dependent, nonlysosomal cysteine proteases present in the cytosol as the inactive proenzyme, procalpain, which translocates to the cell membrane in the presence of Ca^{2+}. One important substrate for calpain is α-spectrin (or fodrin) implicated as a death substrate, which cleaved and played a role in regulating membrane structure, cell shape, and linking the cytoskeleton to the plasma membrane or intracellular vehicles (Martin et al., 1995). LMWF degraded α-spectrin into 150- and 120-kDa fragments. The E-64d calpain inhibitor reduced the production of the 150-kDa fragment; these findings suggest that both calpain and caspases participated in LMWF-induced apoptosis (Zhang et al., 2013).

The proposed mechanism for FCSPs mediated inhibition of proliferation and apoptosis of melanoma cells, and activation of macrophages via membrane receptors, which leads to the production of cytokines that enhance natural killer (NK) cell activation. Activated NK cells release Granzyme

B and perforin through granule exocytosis into the space between NK cells and melanoma cells to initiate caspase cascades in melanoma cells. Granzyme B then initiates apoptosis by triggering the release of mitochondrial cytochrome *c* and apoptosome formation leading to caspase-3 activation, which in turn translocates the nucleus causing DNA fragmentation the distinct morphological change of cells by apoptosis (Ale et al., 2011). The antitumor activity of fucoidan from *F. vesiculosus* was investigated in human colon carcinoma cells. The crude fucoidan, composed predominantly of sulfated fucose, markedly inhibited the growth of HCT-15 cells (human colon carcinoma cells) by several apoptotic events such as DNA fragmentation, chromatin condensation, and increase in the population of sub-G1 cells and mediates apoptosis signaling through mitochondrial pathway (Hyun et al., 2009).

3.2. Role of fucoidan on metastasis, angiogenesis, and signaling mechanism

Metastasis is a leading cause (up to 90%) of cancer-related deaths. Metastasis is a complex series of steps in which cancer cells leave the original tumor site and migrate to other parts of the body via the bloodstream, the lymphatic system, or by direct extension. Malignant cells break away from the primary tumor and attach and degrade proteins that make up the surrounding extracellular matrix (ECM), which separates the tumor from adjoining tissues. By degrading these proteins, cancer cells are able to break the ECM and escape. Matrix metalloproteinases (MMPs) play a key role in tumor metastasis. Fucoidan suppressed MMP-2 activity and protein expression with increasing concentrations of fucoidan in A549 (lung) cancer cells (Lee, Kim, & Kim, 2012). Fucoidan inhibits metastasis by MMP-2 expression and secretion of vascular endothelial growth factor (VEGF) (Ye et al., 2005). The antimetastatic activity of fucoidan was also proven in the animal model of experimental transplanted Lewis lung carcinoma cells (Alekseyenko et al., 2007). Also fucoidan regulates the several cell surface proteins involved in migration and cell adhesion, including integrins, VEGF-1 and -2, P-selectin, and neuropilin-1 (Bachelet et al., 2009; Lake et al., 2006; Rouzet et al., 2011).

New growth in the vascular network is important for the cell proliferation, as well as metastatic spread, of cancer cells depending on an adequate supply of oxygen and nutrients. New blood and lymphatic vessels form through processes called angiogenesis and lymphangiogenesis (Nishida, Yano, Nishida, Kamura, & Kojiro, 2006). Antiangiogenic therapy has become an effective strategy for inhibiting tumor growth. Inhibitors of angiogenesis are attracted target for cancer therapy (Nelson, 1998). Oversulfated fucoidan enhances antitumor and antiangiogenic effects on cancer (Koyanagi, Tanigawa, Nakagawa, Soeda, & Shimeno, 2003). SDF-1 (stromal cell-derived factor-1) is a small cytokine belonging to the chemokine family designated as chemokine (C-X-C motif) ligand 12 (CXCL12). The receptor for this chemokine is CXCR4. Targeting the CXCL12/CXCR4 pathway is a logic strategy in cancer therapy (Arya, Ahmed, Silhi, Williamson, & Patel, 2007; Kryczek, Wei, Keller, Liu, & Zou, 2007). Fucoidan inhibited cell growth more prominently in Huh7 human hepatic cancer cells by suppressing chemotaxin CXCL12 and CXCR4 (Nagamine et al., 2009). The interaction of oversulfated FCSPs with VEGF165 occurred with high affinity and resulted in the formation of highly stable complexes, thereby interfering with the binding of VEGF165 to vascular endothelial growth factor receptor-2 (VEGFR-2). Both native and oversulfated FCSPs were able to suppress neovascularization in mice implanted Sarcoma 180 cells; and that both FCSP types inhibited tumor growth through the prevention of tumor-induced angiogenesis (Koyanagi et al., 2003). Fucoidan inhibited angiogenesis of both human glioblastoma (T98G) and THP1 (acute monocytic leukemia) cells (Lv et al., 2012).

The *ex vivo* angiogenesis assay demonstrated that 100 μg/ml of fucoidan caused significant reduction in microvessel outgrowth and significantly reduced the expression of the angiogenesis factor VEGF-A (Liu et al., 2012). Fucoidan significantly reduces tumor volume and the number of metastatic lung nodules in the 4T1 xenograft model. Fucoidan suppressed *in vitro* cell proliferation, colony formation, and expression of epithelial to mesenchymal transition (EMT) biomarkers, and blocked cell migration and cell invasion (Hsu et al., 2013). The molecular network of transforming growth factor-β (TGF-β) receptors (TGFRs) plays an important role in the regulation of the EMT in cancer cells. Fucoidan decreases TGF-RI and TGF-RII proteins and affects downstream signaling molecules, including Smad2/3 phosphorylation and Smad4 expression. The fucoidan antitumor activity was mediated by ubiquitin-dependent degradation pathway that affects the TGFR/Smad/Snail, Slug, Twist, and EMT axes (Hsu et al., 2013). Mitogen-activated protein kinase pathways are involved in cellular proliferation, differentiation, and apoptosis (Chang & Karin, 2001; Wada & Penninger, 2004). ERK1/2 pathway is involved in the invasive or migratory behavior of a number of malignancies (Smalley, 2003; Suthiphongchai, Promyart, Virochrut, Tohtong, & Wilairat, 2003; Whitmarsh & Davis, 2007). Antimetastatic effect of fucoidan is owing to its inactivation of ERK1/2 pathway in A549 human lung cancer cells. In addition to ERK1/2, the PI3K-Akt signal pathway has been shown to regulate the invasion and metastasis of nonsmall cell lung cancer (NSCLC) as well as the development and progress of various other tumors. PI3K overexpression is highly correlated with the development, invasion, and metastasis of NSCLC (Liao, Wang, Zhang, & Liu, 2006). Fucoidan inhibits the phosphorylation of PI3K-Akt in time- and concentration-dependent manner. PI3K and Akt are well-known upstream regulators of mTOR (mammalian target of rapamycin) signaling pathway in mammalian cells. Fucoidan effectively downregulates the expression of MMP-2 through the inhibitions of PI3K-Akt-mTOR as well as ERK1/2 signaling pathways in A549 human lung cancer cells. Fucoidan significantly inhibited phospho-ERK1/2, p-PI3K, and p-AKt in a concentration-dependent manner with a maximum inhibitory effect at the highest concentration (200 µg/ml). Additionally, 4E-BP1 and p70S6K, two immediate downstream targets of mTOR and indicators of mTOR activity, were also significantly suppressed (Lee et al., 2012). NF-κB and activator protein-1 (AP-1) are transcription factors that regulate the expressions of numerous genes associated with many important biological and pathological processes including cancer. Inhibition of NF-κB and AP-1 results in the suppression of tumor initiation, promotion, and metastasis (Busch, Renaud, Schleussner, Graham, & Markert, 2009; Wu, Wu, & Hu, 2008). Fucoidan inhibits the phosphorylation of Iκ-Bα and increased the total p65 in the cytosolic fraction, and it is decreased in nuclear fraction in lung cancer cells (Lee et al., 2012). Fucoidan exerted a potent inhibitory effect on EGF-induced phosphorylation of EGFR and consequently suppressed the phosphorylation of ERK and JNK. Also, c-fos, c-jun, and AP-1 transcriptional activities were inhibited by fucoidan (Lee, Ermakova, et al., 2008). AP-1 binds to a specific target DNA site in the promoters of several cellular genes and mediates immediate early gene expression involved in a diverse set of transcriptional regulation processes (Hsu, Young, Cmarik, & Colburn, 2000). Fucoidan suppresses the AP-1 activation through inhibition of JunD expression in an HTLV-1-infected T-cell line, thereby inhibiting HTLV-1-infected T-cell proliferation (Haneji et al., 2005). The overview of fucoidan regulation on signaling molecules is shown in Fig. 11.2.

4. CONCLUSIONS

Brown seaweeds contain fucoidans that are complex and heterogeneous structures, but not been very clear until now. Fucoidan has various

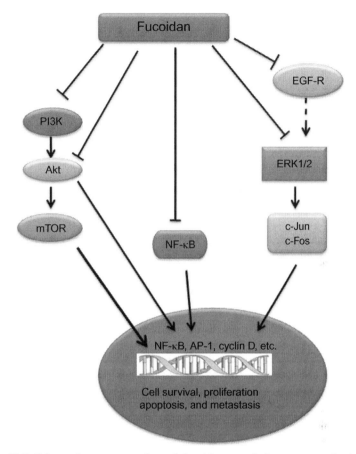

Figure 11.2 Schematic representation of fucoidan regulation on growth signaling molecules.

biological activities including antitumor activity. Fucoidan inhibits cancer cell proliferations by inducing cell cycle arrest and apoptosis, inhibiting metastasis and angiogenesis, and regulating signaling molecule. Therefore, fucoidan or LMWF could be a useful target for cancer therapy.

ACKNOWLEDGMENT

This study was supported by a grant from Marine Bioprocess Research Center of the Marine Bio 21 Project funded by the Ministry of Land, Transport and Maritime, Republic of Korea.

REFERENCES

Aisa, Y., Miyakawa, Y., Nakazato, T., Shibata, H., Saito, K., Ikeda, Y., et al. (2005). Fucoidan induces apoptosis of human HS Sultan cells accompanied by activation of caspase-3 and down-regulation of ERK pathways. *American Journal of Hematology*, 78(1), 7–14.

Ale, M. T., Maruyama, H., Tamauchi, H., Mikkelsen, J. D., & Meyer, A. S. (2011). Fucoidan from *Sargassum sp.* and *Fucus vesiculosus* reduces cell viability of lung carcinoma and melanoma cells *in vitro* and activates natural killer cells in mice *in vivo*. *International Journal of Biological Macromolecules, 49*(3), 331–336.

Alekseyenko, T., Zhanayeva, S. Y., Venediktova, A., Zvyagintseva, T., Kuznetsova, T., Besednova, N., et al. (2007). Antitumor and antimetastatic activity of fucoidan, a sulfated polysaccharide isolated from the Okhotsk Sea *Fucus evanescens* brown alga. *Bulletin of Experimental Biology and Medicine, 143*(6), 730–732.

Anastyuk, S. D., Imbs, T. I., Shevchenko, N. M., Dmitrenok, P. S., & Zvyagintseva, T. N. (2012). Compositional and structural mass spectrometric analysis of fucoidan preparations from *Costaria costata*, extracted from alga at different life-stages. *Carbohydrate Polymers, 90*, 993–1002.

Arya, M., Ahmed, H., Silhi, N., Williamson, M., & Patel, H. R. (2007). Clinical importance and therapeutic implications of the pivotal CXCL12–CXCR4 (chemokine ligand-receptor) interaction in cancer cell migration. *Tumor Biology, 28*(3), 123–131.

Ashkenazi, A., & Dixit, V. M. (1998). Death receptors: Signaling and modulation. *Science, 281*(5381), 1305–1308.

Athukorala, Y., Jung, W.-K., Vasanthan, T., & Jeon, Y.-J. (2006). An anticoagulative polysaccharide from an enzymatic hydrolysate of *Ecklonia cava*. *Carbohydrate Polymers, 66*(2), 184–191.

Bachelet, L., Bertholon, I., Lavigne, D., Vassy, R., Jandrot-Perrus, M., Chaubet, F., et al. (2009). Affinity of low molecular weight fucoidan for P-selectin triggers its binding to activated human platelets. *Biochimica et Biophysica Acta, 1790*(2), 141–146.

Banafa, A. M., & Roshan, S. (2013). Fucoidan induces G1 phase arrest and apoptosis through caspases-dependent pathway and ROS induction in human breast cancer MCF-7 cells. *Journal of Huazhong University of Science and Technology, 33*(5), 717–724.

Berteau, O., & Mulloy, B. (2003). Sulfated fucans, fresh perspectives: Structures, functions, and biological properties of sulfated fucans and an overview of enzymes active toward this class of polysaccharide. *Glycobiology, 13*(6), 29–40.

Bhadury, P., Mohammad, B. T., & Wright, P. C. (2006). The current status of natural products from marine fungi and their potential as anti-infective agents. *Journal of Industrial Microbiology & Biotechnology, 33*(5), 325–337.

Bilan, M. I., & Usov, A. I. (2009). Structural analysis of fucoidans. *ChemInform, 40*, 34.

Boisson-Vidal, C., Haroun, F., Ellouali, M., Blondin, C., Fischer, A., de Agostini, A., et al. (1995). Biological activities of polysaccharides from marine algae. *Drugs of the Future, 20*(12), 1237–1250.

Busch, S., Renaud, S. J., Schleussner, E., Graham, C. H., & Markert, U. R. (2009). mTOR mediates human trophoblast invasion through regulation of matrix-remodeling enzymes and is associated with serine phosphorylation of STAT3. *Experimental Cell Research, 315*(10), 1724–1733.

Chandini, S. K., Ganesan, P., Suresh, P., & Bhaskar, N. (2008). Seaweeds as a source of nutritionally beneficial compounds—A review. *Journal of Food Science and Technology, 45*(1), 1–13.

Chang, L., & Karin, M. (2001). Mammalian MAP kinase signalling cascades. *Nature, 410*(6824), 37–40.

Changotade, S., Korb, G., Bassil, J., Barroukh, B., Willig, C., Colliec Jouault, S., et al. (2008). Potential effects of a low molecular weight fucoidan extracted from brown algae on bone biomaterial osteoconductive properties. *Journal of Biomedical Materials Research. Part A, 87*(3), 666–675.

Chevolot, L., Mulloy, B., Ratiskol, J., Foucault, A., & Colliec-Jouault, S. (2001). A disaccharide repeat unit is the major structure in fucoidans from two species of brown algae. *Carbohydrate Research, 330*(4), 529–535.

Costa, L., Fidelis, G., Cordeiro, S., Oliveira, R., Sabry, D., Câmara, R., et al. (2010). Biological activities of sulfated polysaccharides from tropical seaweeds. *Biomedicine & Pharmacotherapy*, 64(1), 21–28.

Costantino, V., Fattorusso, E., Menna, M., & Taglialatela-Scafati, O. (2004). Chemical diversity of bioactive marine natural products: An illustrative case study. *Current Medicinal Chemistry*, 11(13), 1671–1692.

Cumashi, A., Ushakova, N. A., Preobrazhenskaya, M. E., D'Incecco, A., Piccoli, A., Totani, L., et al. (2007). A comparative study of the anti-inflammatory, anticoagulant, antiangiogenic, and antiadhesive activities of nine different fucoidans from brown seaweeds. *Glycobiology*, 17(5), 541–552.

Desagher, S., Osen-Sand, A., Nichols, A., Eskes, R., Montessuit, S., Lauper, S., et al. (1999). Bid-induced conformational change of Bax is responsible for mitochondrial cytochrome c release during apoptosis. *The Journal of Cell Biology*, 144(5), 891–901.

Duarte, M. E., Cardoso, M. A., Noseda, M. D., & Cerezo, A. S. (2001). Structural studies on fucoidans from the brown seaweed *Sargassum stenophyllum*. *Carbohydrate Research*, 333(4), 281–293.

Duprez, L., Wirawan, E., Berghe, T. V., & Vandenabeele, P. (2009). Major cell death pathways at a glance. *Microbes and Infection*, 11(13), 1050–1062.

Ermakova, S., Sokolova, R., Kim, S.-M., Um, B.-H., Isakov, V., & Zvyagintseva, T. (2011). Fucoidans from brown seaweeds *Sargassum hornery*, *Eclonia cava*, *Costaria costata*: Structural characteristics and anticancer activity. *Applied Biochemistry and Biotechnology*, 164(6), 841–850.

Eskes, R., Desagher, S., Antonsson, B., & Martinou, J.-C. (2000). Bid induces the oligomerization and insertion of Bax into the outer mitochondrial membrane. *Molecular and Cellular Biology*, 20(3), 929–935.

Fukahori, S., Yano, H., Akiba, J., Ogasawara, S., Momosaki, S., Sanada, S., et al. (2008). Fucoidan, a major component of brown seaweed, prohibits the growth of human cancer cell lines *in vitro*. *Molecular Medicine Reports*, 1, 537–542.

Gideon, T. P., & Rengasamy, R. (2008). Toxicological evaluation of fucoidan from *Cladosiphon okamuranus*. *Journal of Medicinal Food*, 11(4), 638–642.

Grossi, F., Kubota, K., Cappuzzo, F., de Marinis, F., Gridelli, C., Aita, M., et al. (2010). Future scenarios for the treatment of advanced non-small cell lung cancer: Focus on taxane-containing regimens. *The Oncologist*, 15(10), 1102–1112.

Haneji, K., Matsuda, T., Tomita, M., Kawakami, H., Ohshiro, K., Uchihara, J.-N., et al. (2005). Fucoidan extracted from *Cladosiphon okamuranus* Tokida induces apoptosis of human T-cell leukemia virus type 1-infected T-cell lines and primary adult T-cell leukemia cells. *Nutrition and Cancer*, 52(2), 189–201.

Haroun-Bouhedja, F., Ellouali, M., Sinquin, C., & Boisson-Vidal, C. (2000). Relationship between sulfate groups and biological activities of fucans. *Thrombosis Research*, 100(5), 453–459.

Hayakawa, K., & Nagamine, T. (2009). Effect of fucoidan on the biotinidase kinetics in human hepatocellular carcinoma. *Anticancer Research*, 29(4), 1211–1217.

Hengartner, M. O. (2000). The biochemistry of apoptosis. *Nature*, 407(6805), 770–776.

Holdt, S. L., & Kraan, S. (2011). Bioactive compounds in seaweed: Functional food applications and legislation. *Journal of Applied Phycology*, 23(3), 543–597.

Hsu, H.-Y., Lin, T.-Y., Hwang, P.-A., Chen, R.-H., Tsao, S.-M., & Hsu, J. (2013). Fucoidan induces changes in the epithelial-mesenchymal transition and decreases metastasis by enhancing ubiquitin-dependent TGFβ receptor degradation in breast cancer. *Carcinogenesis*, 34, 874–884.

Hsu, T.-C., Young, M. R., Cmarik, J., & Colburn, N. H. (2000). Activator protein 1 (AP-1)- and nuclear factor κB (NF-κB)-dependent transcriptional events in carcinogenesis. *Free Radical Biology and Medicine*, 28(9), 1338–1348.

Hyun, J.-H., Kim, S.-C., Kang, J.-I., Kim, M.-K., Boo, H.-J., Kwon, J.-M., et al. (2009). Apoptosis inducing activity of fucoidan in HCT-15 colon carcinoma cells. *Biological and Pharmaceutical Bulletin*, *32*(10), 1760–1764.

Jensen, A. (1956). *Component sugars of some common brown algae*. Blindern, Oslo: Norwegian Institute of Seaweed Research, Akademisk Trykningssentral, Report 9.

Jiao, G., Yu, G., Zhang, J., & Ewart, H. S. (2011). Chemical structures and bioactivities of sulfated polysaccharides from marine algae. *Marine Drugs*, *9*(2), 196–223.

Jin, J. O., Song, M. G., Kim, Y. N., Park, J. I., & Kwak, J. Y. (2010). The mechanism of fucoidan induced apoptosis in leukemic cells: Involvement of ERK1/2, JNK, glutathione, and nitric oxide. *Molecular Carcinogenesis*, *49*(8), 771–782.

Johnstone, R. W., Ruefli, A. A., & Lowe, S. W. (2002). Apoptosis—A link between cancer genetics and chemotherapy. *Cell*, *108*(2), 153–164.

Kaufmann, S. H., Desnoyers, S., Ottaviano, Y., Davidson, N. E., & Poirier, G. G. (1993). Specific proteolytic cleavage of poly(ADP-ribose) polymerase: An early marker of chemotherapy-induced apoptosis. *Cancer Research*, *53*(17), 3976–3985.

Kim, E., Park, S., Lee, J.-Y., & Park, J. (2010). Fucoidan present in brown algae induces apoptosis of human colon cancer cells. *BMC Gastroenterology*, *10*(1), 96.

Koyanagi, S., Tanigawa, N., Nakagawa, H., Soeda, S., & Shimeno, H. (2003). Oversulfation of fucoidan enhances its anti-angiogenic and antitumor activities. *Biochemical Pharmacology*, *65*(2), 173–179.

Kryczek, I., Wei, S., Keller, E., Liu, R., & Zou, W. (2007). Stroma-derived factor (SDF-1/CXCL12) and human tumor pathogenesis. *American Journal of Physiology. Cell Physiology*, *292*(3), C987–C995.

Kylin, H. (1913). Zur Biochemie der Meeresalgen. *Hoppe-Seyler's Zeitschrift für Physiologische Chemie*, *83*(3), 171–197.

Lake, A. C., Vassy, R., Di Benedetto, M., Lavigne, D., Le Visage, C., Perret, G. Y., et al. (2006). Low molecular weight fucoidan increases VEGF165-induced endothelial cell migration by enhancing VEGF165 binding to VEGFR-2 and NRP1. *Journal of Biological Chemistry*, *281*(49), 37844–37852.

Lee, S.-H., Athukorala, Y., Lee, J.-S., & Jeon, Y.-J. (2008). Simple separation of anticoagulant sulfated galactan from marine red algae. *Journal of Applied Phycology*, *20*(6), 1053–1059.

Lee, N. Y., Ermakova, S. P., Zvyagintseva, T. N., Kang, K. W., Dong, Z., & Choi, H. S. (2008). Inhibitory effects of fucoidan on activation of epidermal growth factor receptor and cell transformation in JB6 Cl41 cells. *Food and Chemical Toxicology*, *46*(5), 1793–1800.

Lee, H., Kim, J.-S., & Kim, E. (2012). Fucoidan from seaweed *Fucus vesiculosus* inhibits migration and invasion of human lung cancer cell via PI3K-Akt-mTOR pathways. *PLoS One*, *7*(11), e50624.

Li, J., & Yuan, J. (2008). Caspases in apoptosis and beyond. *Oncogene*, *27*(48), 6194–6206.

Liao, D., Wang, L., Zhang, X., & Liu, M. (2006). Expression and significance of PTEN/PI3K signal transduction-related proteins in non-small cell lung cancer. *Ai Zheng Aizheng Chinese Journal of Cancer*, *25*(10), 1238.

Liu, F., Wang, J., Chang, A. K., Liu, B., Yang, L., Li, Q., et al. (2012). Fucoidan extract derived from Undaria pinnatifida inhibits angiogenesis by human umbilical vein endothelial cells. *Phytomedicine*, *19*, 797–803.

Lockshin, R. A., & Zakeri, Z. (2007). Cell death in health and disease. *Journal of Cellular and Molecular Medicine*, *11*(6), 1214–1224.

Lv, Y., Song, Q., Shao, Q., Gao, W., Mao, H., Lou, H., et al. (2012). Comparison of the effects of marchantin C and fucoidan on sFlt-1 and angiogenesis in glioma microenvironment. *Journal of Pharmacy and Pharmacology*, *64*(4), 604–609.

Makarenkova, I., Deriabin, P., L'vov, D., Zviagintseva, T., & Besednova, N. (2010). Antiviral activity of sulfated polysaccharide from the brown algae *Laminaria japonica* against avian influenza A (H5N1) virus infection in the cultured cells. *Voprosy Virusologii*, 55(1), 41.

Martin, S., Reutelingsperger, C., McGahon, A. J., Rader, J. A., Van Schie, R., LaFace, D. M., et al. (1995). Early redistribution of plasma membrane phosphatidylserine is a general feature of apoptosis regardless of the initiating stimulus: Inhibition by overexpression of Bcl-2 and Abl. *The Journal of Experimental Medicine*, 182(5), 1545–1556.

Maruyama, H., Tamauchi, H., Iizuka, M., & Nakano, T. (2006). The role of NK cells in antitumor activity of dietary fucoidan from *Undaria pinnatifida sporophylls* (Mekabu). *Planta Medica*, 72(15), 1415.

Matou, S., Helley, D., Chabut, D., Bros, A., & Fischer, A.-M. (2002). Effect of fucoidan on fibroblast growth factor-2 induced angiogenesis *in vitro*. *Thrombosis Research*, 106(4), 213–221.

Matsubara, K., Xue, C., Zhao, X., Mori, M., Sugawara, T., & Hirata, T. (2005). Effects of middle molecular weight fucoidans on in vitro and ex vivo angiogenesis of endothelial cells. *International Journal of Molecular Medicine*, 15(4), 695–700.

Mestechkina, N., & Shcherbukhin, V. (2010). Sulfated polysaccharides and their anticoagulant activity: A review. *Applied Biochemistry and Microbiology*, 46(3), 267–273.

Murata, M., & Nakazoe, J. I. (2001). Production and use of marine algae in Japan. *Japan Agricultural Research Quarterly*, 35(4), 281–290.

Nagamine, T., Hayakawa, K., Kusakabe, T., Takada, H., Nakazato, K., Hisanaga, E., et al. (2009). Inhibitory effect of fucoidan on Huh7 hepatoma cells through downregulation of CXCL12. *Nutrition and Cancer*, 61(3), 340–347.

Nelson, N. J. (1998). Inhibitors of angiogenesis enter phase III testing. *Journal of the National Cancer Institute*, 90(13), 960–963.

Nishida, N., Yano, H., Nishida, T., Kamura, T., & Kojiro, M. (2006). Angiogenesis in cancer. *Vascular Health and Risk Management*, 2(3), 213–219.

Norbury, C., & Nurse, P. (1992). Animal cell cycles and their control. *Annual Review of Biochemistry*, 61(1), 441–468.

Patankar, M. S., Oehninger, S., Barnett, T., Williams, R., & Clark, G. (1993). A revised structure for fucoidan may explain some of its biological activities. *Journal of Biological Chemistry*, 268(29), 21770–21776.

Patel, S. (2012). Therapeutic importance of sulfated polysaccharides from seaweeds: updating the recent findings. *Biotech*, 3(2), 1–15.

Percival, E., & McDowell, R. H. (1967). *Chemistry and enzymology of marine algal polysaccharides*. London: Academic Press, p. 219.

Renn, D. (1997). Biotechnology and the red seaweed polysaccharide industry: Status, needs and prospects. *Trends in Biotechnology*, 15(1), 9–14.

Riou, D., Colliec-Jouault, S., Pinczon du Sel, D., Bosch, S., Siavoshian, S., Le Bert, V., et al. (1996). Antitumor and antiproliferative effects of a fucan extracted from ascophyllum nodosum against a non-small-cell bronchopulmonary carcinoma line. *Anticancer Research*, 16(3A), 1213.

Rouzet, F., Bachelet-Violette, L., Alsac, J.-M., Suzuki, M., Meulemans, A., Louedec, L., et al. (2011). Radiolabeled fucoidan as a p-selectin targeting agent for in vivo imaging of platelet-rich thrombus and endothelial activation. *Journal of Nuclear Medicine*, 52(9), 1433–1440.

Schneider, U., Stipper, A., & Besserer, J. (2010). Dose–response relationship for lung cancer induction at radiotherapy dose. *Zeitschrift für Medizinische Physik*, 20(3), 206–214.

Semenov, A., Mazurov, A., Preobrazhenskaia, M., Ushakova, N., Mikhaĭlov, V., Berman, A., et al. (1998). Sulfated polysaccharides as inhibitors of receptor activity of P-selectin and P-selectin-dependent inflammation. *Voprosy meditsinskoĭ khimii*, *44*(2), 135.

Senni, K., Gueniche, F., Foucault-Bertaud, A., Igondjo-Tchen, S., Fioretti, F., Colliec-Jouault, S., et al. (2006). Fucoidan a sulfated polysaccharide from brown algae is a potent modulator of connective tissue proteolysis. *Archives of Biochemistry and Biophysics*, *445*(1), 56–64.

Senthilkumar, K., Manivasagan, P., Venkatesana, J., & Kim, S.-K. (2013). Brown seaweed fucoidan: Biological activity and apoptosis, growth signaling mechanism in cancer. *International Journal of Biological Macromolecules*, *60*, 366–374.

Smalley, K. S. (2003). A pivotal role for ERK in the oncogenic behaviour of malignant melanoma? *International Journal of Cancer*, *104*(5), 527–532.

Sprick, M. R., & Walczak, H. (2004). The interplay between the Bcl-2 family and death receptor-mediated apoptosis. *Biochimica et Biophysica Acta*, *1644*(2), 125–132.

Suthiphongchai, T., Promyart, P., Virochrut, S., Tohtong, R., & Wilairat, P. (2003). Involvement of ERK1⁄2 in invasiveness and metastatic development of rat prostatic adenocarcinoma. *Oncology Research Featuring Preclinical and Clinical Cancer Therapeutics*, *13*(5), 253–259.

Tako, M., Yoza, E., & Tohma, S. (2000). Chemical characterization of acetyl fucoidan and alginate from commercially cultured Cladosiphon okamuranus. *Botanica Marina*, *43*(4), 393–398.

Teruya, T., Konishi, T., Uechi, S., Tamaki, H., & Tako, M. (2007). Anti-proliferative activity of oversulfated fucoidan from commercially cultured *Cladosiphon okamuranus* TOKIDA in U937 cells. *International Journal of Biological Macromolecules*, *41*(3), 221–226.

Veena, C. K., Josephine, A., Preetha, S. P., Varalakshmi, P., & Sundarapandiyan, R. (2006). Renal peroxidative changes mediated by oxalate: The protective role of fucoidan. *Life Sciences*, *79*(19), 1789–1795.

Vishchuk, O. S., Ermakova, S. P., & Zvyagintseva, T. N. (2011). Sulfated polysaccharides from brown seaweeds *Saccharina japonica* and *Undaria pinnatifida*: Isolation, structural characteristics and antitumor activity. *Carbohydrate Research*, *346*(17), 2769–2776.

Wada, T., & Penninger, J. M. (2004). Mitogen-activated protein kinases in apoptosis regulation. *Oncogene*, *23*(16), 2838–2849.

Wang, J., Zhang, Q., Zhang, Z., Song, H., & Li, P. (2010). Potential antioxidant and anticoagulant capacity of low molecular weight fucoidan fractions extracted from *Laminaria japonica*. *International Journal of Biological Macromolecules*, *46*(1), 6–12.

Whitmarsh, A., & Davis, R. (2007). Role of mitogen-activated protein kinase kinase 4 in cancer. *Oncogene*, *26*(22), 3172–3184.

Wu, W.-S., Wu, J.-R., & Hu, C.-T. (2008). Signal cross talks for sustained MAPK activation and cell migration: The potential role of reactive oxygen species. *Cancer and Metastasis Reviews*, *27*(2), 303–314.

Yamasaki-Miyamoto, Y., Yamasaki, M., Tachibana, H., & Yamada, K. (2009). Fucoidan induces apoptosis through activation of caspase-8 on human breast cancer MCF-7 cells. *Journal of Agricultural and Food Chemistry*, *57*(18), 8677–8682.

Ye, J., Li, Y., Teruya, K., Katakura, Y., Ichikawa, A., Eto, H., et al. (2005). Enzyme-digested fucoidan extracts derived from seaweed Mozuku of *Cladosiphon novae-caledoniae kylin* inhibit invasion and angiogenesis of tumor cells. *Cytotechnology*, *47*(1–3), 117–126.

Zhang, Q., Li, N., Zhao, T., Qi, H., Xu, Z., & Li, Z. (2005). Fucoidan inhibits the development of proteinuria in active Heymann nephritis. *Phytotherapy Research*, *19*(1), 50–53.

Zhang, Z., Teruya, K., Eto, H., & Shirahata, S. (2011). Fucoidan extract induces apoptosis in MCF-7 cells via a mechanism involving the ROS-dependent JNK activation and mitochondria-mediated pathways. *PLoS One*, *6*(11), 1–14.

Zhang, Z., Teruya, K., Eto, H., & Shirahata, S. (2013). Induction of apoptosis by low molecular weight fucoidan through calcium and caspase dependent mitochondrial pathways in MDA-MB-231 breast cancer cells. *Bioscience, Biotechnology, and Biochemistry*, 77, 235–242.

Zhu, Z., Zhang, Q., Chen, L., Ren, S., Xu, P., Tang, Y., et al. (2010). Higher specificity of the activity of low molecular weight fucoidan for thrombin-induced platelet aggregation. *Thrombosis Research*, 125(5), 419–426.

Zhuang, C., Itoh, H., Mizuno, T., & Ito, H. (1995). Antitumor active fucoidan from the brown seaweed, umitoranoo (*Sargassum thunbergii*). *Bioscience, Biotechnology, and Biochemistry*, 59(4), 563.

CHAPTER TWELVE

Anticancer Effects of Chitin and Chitosan Derivatives

Mustafa Zafer Karagozlu*, Se-Kwon Kim[†,1]

*Marine Bioprocess Research Center, Pukyong National University, Busan, South Korea
[†]Department of Marine-bio Convergence Science, Specialized Graduate School Science and Technology Convergence, Marine Bioprocess Research Center, Pukyong National University, Busan, South Korea
[1]Corresponding author: e-mail address: sknkim@pknu.ac.kr

Contents

1. Introduction	215
2. Anticancer Activity as a Therapeutic Agent	217
3. Anticancer Activity as a Carrier	220
4. Conclusion	222
References	222

Abstract

Despite considerable progress in medical research, cancer is still one of the high-ranking causes of death in the world. It is the second most common cause of death due to disease after heart disease, and according to World Health Organization it will be the cause of death for more than 10 million people in 2020; therefore, one of the main research goals for researchers investigating new anticancer agents. But the major complication for the cancer cure without surgeries is side effects. Especially, cytotoxic anticancer chemotherapeutic agents generally produce severe side effects, while reducing host resistance to cancer and infections. Therefore, it is important to find new, powerful anticancer agents that are highly effective, biodegradable, and biocompatible. Chitin and chitosan are biopolymers which have unique structural possibilities for chemical and mechanical modifications to generate novel properties, functions. These biopolymers are biocompatible, biodegradable, and nontoxic, and their chemical properties allow them to be easily processed into gels, sponges, membranes, beads, and scaffolds forms also. Due to their unique properties, they are excellent candidates for cancer cure or cancer diagnosis.

1. INTRODUCTION

Chitin is a natural polysaccharide which has been firstly identified in 1821. Henri Braconnot, who is the director of the botanical garden in France, observed a material in mushrooms which did not dissolve in sulfuric acid. Braconnot named it as fungine. In the late 1830s, it was isolated from insects, and in 1859, chitosan, a derivative of chitin, was produced (Nicol, 1991).

Chitin is synthesized by an enormous number of living organisms, and it is the second most abundant polymer after cellulose. Especially, it occurs in nature as ordered crystalline microfibril forming structural components in the exoskeleton of arthropods or in the cell walls of fungi and yeast. It is also produced by a number of other living organisms in the lower plant and animal kingdoms, serving in many functions where reinforcement and strength are required.

Studies on chitin and chitin derivatives have been intensified since 1990 because these polysaccharides show excellent biological properties such as biodegradation in the human body (Sashiwa, Saimoto, Shigemasa, Ogawa, & Tokura, 1990; Shigemasa, Saito, Sashiwa, & Saimoto, 1994), immunological (Mori et al., 1997; Nishimura et al., 1984), antibacterial (Tanigawa, Tanaka, Sashiwa, Saimoto, & Shigemasa, 1992; Tokura, Ueno, Miyazaki, & Nishi, 1997), and wound-healing activity (Khnor & Lim, 2003; Kweon, Song, & Park, 2003; Okamoto et al., 1993).

On the other hand, chitosan is a natural nontoxic heteropolysaccharide composed of β-1,4-linked-D-glucosamine (GlcN) and N-acetyl-D-glucosamine (GlcNAc) in varying proportions. These polysaccharides have been widely studied and applied in different fields. Even chitosan is a derivative of chitin; it has its own unique functions such as support material for gene delivery (Sato, Ishii, & Okahata, 2001), cell culture (Mao, Liu, Yin, & Yao, 2003), and tissue engineering (Gingras, Paradis, & Berthod, 2003; Wang, Lin, Wang, & Hsieh, 2003). Especially in the biomedical and pharmaceutical industries, chitosan and chitosan derivatives have been widely used due to their various biological functions to their various biological functions such as antimicrobial (Kong, Kim, Ahn, Byun, & Kim, 2010), antibacterial (Jeon, Park, & Kim, 2001), antioxidant (Park, Je, & Kim, 2004), and immunostimulating (Huang, Mendis, Rajapakse, & Kim, 2006) effects. Besides, they have antitumor (Seo et al., 2000), antidiabetic (Karadeniz, Artan, Kim, & Kim, 2008), and antiviral (Artan, Karadeniz, Karagozlu, Kim, & Kim, 2010) activities. Unfortunately, poor solubility of chitosan is principal limiting factor for its wide application. Thereupon recent studies on chitosan have attracted interest in converting it to more soluble form like chitooligosaccharides (COS). COS are the degraded oligomers of chitosan, which can be obtained by either chemical (Defaye & Guillot, 1994; Horowitz, Roseman, & Blumenthal, 1957; Tsukada &

Inoue, 1981) or enzymatic (Izume & Ohtakara, 1987) hydrolysis of chitosan. It has been reported that lower oligomers of chitosan are not only water-soluble but also exhibit versatile biological activities similar to chitosan (Qin, Du, Xiao, Li, & Gao, 2002). The biological activity of COS is known to depend on their structure and molecular weight (Hahn, 1996). Beside, investigators mentioned that properties of COS, such as degrees of polymerization, degrees of acetylation, charge distribution, and nature of chemical modification to the molecule strongly influence its observed biological activities (Muzzarelli, 1997).

2. ANTICANCER ACTIVITY AS A THERAPEUTIC AGENT

A main goal of cancer research is to completely prevent recurrence following surgery and to increase life time of the patient. But the major problem of the cure without surgeries is side effects. Especially, cytotoxic anticancer chemotherapeutic agents generally produce severe side effects, while reducing host resistance to cancer and infections. Therefore, it is important to find new, powerful anticancer agents that are highly effective, biodegradable, and biocompatible. Chitosan and chitosan derivative anticancer agents are known to be favorable pharmaceutical materials because of their biocompatible and biodegradable properties (Felt, Buri, & Gurny, 1998).

Several chitin derivatives were investigated for their antitumor activity (Murata, Saiki, Matsuno, Tokura, & Azumo, 1990). Murata et al. (1990) reported that 6-O-sulfated chitin significantly inhibited the lung tumor colonization in proportion to the degree of sulfation. Furthermore, 6-O-sulfated carboxymethylated chitin (SCM-chitin) with a high degree of sulfation caused a marked decreased of number of lung tumor colonies in the spontaneous lung metastasis model. SCM-chitin also significantly inhibited the arrest of B16Bl6 cells in lungs after co-injection with radiolabeled tumor cells.

A large amount of literature exists regarding the effects of antitumor activities of chitosan and its derivatives (Suzuki et al., 1986; Tokoro et al., 1988). Suzuki et al. (1986) found that N-acetyl chitosan oligomer, particularly the hexane and heptamer, display notable antitumor activity against sarcoma 180 solid tumors in BALB/C mice as well as in MM-46 solid tumor implanted in C3H/HC mice. These results indicated that the effect was not by direct cytodial action on the tumor cells and was indeed host-mediated. Tokoro et al. (1988) showed that hexameric chitosan oligomer had growth-inhibitory effect against Meth-A solid tumor transplanted into BALB/C mice. The antitumor mechanism was assumed to be involved in increased production of lymphokines including interleukins 1 and 2, sequentially, leading to manifestation of antitumor effect through proliferation of cytolytic T-lymphocytes. In addition, the antitumor activity of low-molecular-weight chitosan (LMWC) with the higher than hexamer was investigated. Qin et al. (2002) reported that LMWC was prepared by enzymatic hydrolysis using cellulose and hemicellulose and investigated the inhibition of growth of sarcoma 180 tumor cells in mice. Maeda and Kimura (2004) prepared various molecular weights such as 21, 46, and 120 kDa chitosans by enzymatic hydrolysis and examined the antitumor activity in sarcoma 180-bearing mice. The antitumor activity of various molecular weights of chitosans showed that 21 kDa chitosan significantly reduced the tumor growth and final tumor weight. Moreover, 21 and 46 kDa chitosans enhanced the natural killer activity in intestinal intraphelial lymphocytes or splenic lymphocytes. Harish Prashanth and Tharanathan (2005) generated LMWC by depolymerization induced by potassium persulfate under nitrogen atmosphere. Moreover, Jeon et al. (2001) also carried out a study to identify the correlation between molecular weight of COS and their antitumor activity. In their research, different molecular weight COS have been prepared by UF membrane reactor system. The researchers suggested that, medium-molecular-weight molecular COS ranging 1.5–5.5 kDa could effectively inhibit the growth of sarcoma 180 solid (S180) or Uterine cervix carcinoma No. 14 (U14) tumor in BALB/C mice. Hasegawa, Yagi, Iwakawa, and Hirai (2001) reported the growth-inhibitory effect of chitosan on bladder tumor cells. They observed DNA fragmentation, which is characteristic of apoptosis, and elevated caspase-3-like activity in chitosan-treated cancer cells. In addition, modified chitosans were reported to display the growth-inhibitory effect on tumor cells (Sirica & Woodman, 1971), and this property was employed by Ouchi, Inosaka, Banba, and Ohya (1992) by conjugating chitosan or chitosanaminooligosaccharide to 5-fluorouracil (5FU) in order to provide a macromolecular system with strong antitumor activity and reduced side effects. Indeed, the strong antitumor activity exhibited by 5FU is accompanied by undesirable side effects. *In vivo* studies demonstrated that chitosan-5FU conjugate exhibited a strong survival effect against lymphocytic leukemia in mice. Furthermore, chitosan-5FU and COS-5FU conjugates showed remarkable growth-inhibitory effects on Met-A fibrosarcoma and MH-134Y hepatoma. Both conjugates displayed no acute toxicity, even in high-dose ranges. Therefore, they reported that chitosan-5FU and COS-5FU are expected to act clinically as macromolecular prodrugs of 5FU.

Furthermore, studies on antitumor activity of chitosan and COS revealed that partially deacetylated chitin and carboxymethyl chitin with an adequate degree of substitution were effective toward controlling various tumor cells (Nishimura et al., 1984). Unlike many other biological molecules, COS could exert their biological activities following oral administration and effects are more or less similar to those of intraperitoneal injection. Moreover, Qin et al. (2002) have demonstrated that water-soluble COS prepared with a mixture of tetramer and pentamer could inhibit growth of S180 tumor cells in mice after oral and intraperitoneal administration. Therefore, COS and their N-acetylated analogues that are soluble in basic physiologic environments could be considered good candidates to develop potential nutraceuticals.

The antitumor mechanism of these COS was probably related to their induction of T-cell proliferation to produce the tumor-inhibitory effects. Through analysis of the splenic cell changes in cancerous mice, Suzuki et al. (1986) proved that the antitumor mechanism of COS is to enhance acquired immunity by accelerating T-cell differentiation to increase cytotoxicity and maintain T-cell activity. Besides, *in vitro* researches demonstrated that charge properties of the chitosan are also important for anticancer activity. Karagozlu, Karadeniz, Kong, and Kim (2012) and Huang et al. (2006) studied the anticancer activities of differently charged COS derivatives using four cancer cell lines: HeLa, Hep3B, SW480, and AGS. Neutral red and MTT cell-viability studies suggested that highly charged COS derivatives could significantly reduce cancer cell viability, regardless of their positive or negative charge. Furthermore, fluorescence microscopic observations and Western blotting studies confirmed that the anticancer effect of these highly charged COS derivatives was triggering off intrinsic apoptotic pathway.

Laminin are basal proteins in basal lamina and are known to correlate with metastasis of tumor cells. A peptide containing the Tyr-Ile-Gly-SerArg (YIGSR) sequence, corresponding to a partial sequence of laminin, inhibited angiogenesis and thus depressed tumor growth. Nishiyama et al. (2000) prepared YIGSR-chitosan conjugate and assayed antimetastatic activity. The conjugate proved to have higher inhibitory activity against experimental lung metastasis of B16BL6 melanoma cells in mice than did the parent peptide.

Kong et al. (2010) also investigated the MMP inhibition of chitin, water-soluble chitosan, and their carboxymethylated derivatives. In the research, chitosan and chitin, carboxymethyl-chitosan (CM-chitosan) and carboxymethyl-chitin (CM-chitin), were synthesized by means of carboxymethylation reaction. Their antioxidative and matrix metalloproteinase-2 and -9 inhibitory effects were investigated in HT1080 human fibrosarcoma cells. The research suggests CM-chitosan and CM-chitin as potent antioxidants and MMP inhibitor via alleviations of radical-induced oxidative damage.

3. ANTICANCER ACTIVITY AS A CARRIER

Chitosan and chitin are also used as drug carriers to provide anticancer and antitumor chemotherapy which can improve drug absorption, stabilize drug components to increase drug targeting, and enhance drug release. As a gene carrier, chitosan can be used for DNA protection and effect the expression period of genes. It has been reported that the conjugates of some kinds of anticancer agents with chitin and chitosan derivatives display good anticancer effects with a decrease in side effects over the original form due to a predominant distribution in the cancer tissue and a gradual release of free drug from the conjugates. For instance, doxorubicin is one of the most used anticancer agents which can load in various polymeric or natural hydrogels (Han et al., 2008; Obara et al., 2005). Cho, Park, Jeong, and Yoo (2009) prepared doxorubicin hydrogel containing COS-DOX to obtain sustained-release profiles of doxorubicin from thermo-responsive and photo-crosslinkable hydrogels and examined its anticancer activity on human lung cancer adenocarcinoma cell line *in vitro* and *in vivo*. The research demonstrated that released fraction composed of doxorubicin and chitosan-doxorubicin oligomers showed comparable *in vitro* cytotoxicity to free doxorubicin. Besides, doxorubicin hydrogels containing chitosan-doxorubicin conjugates showed superior *in vivo* anticancer effects in human solid tumors compared to free doxorubicin or hydrogel containing free doxorubicin after 3 weeks.

Moreover, doxifluridine and 1-β-D-,-D-Arabinofuranosylcytosine (Ara-C) are typical time-dependent antitumor agents. But the major problem was the large doses of Ara-C required because of its resistance in the body. It can be quickly eliminated or inactivated (Aoshima, Tsukagoshi, Sakurai, Oh-ishi, & Ishida, 1976). Thus, various derivatives of Ara-C have been developed in attempts to improve efficacy. One of the derivatives of Ara-C was modified with cytidine deaminase. This modification catalyzes the transformation from an amino group to a hydroxyl group (Onishi, Pithayanukul, & Nagai, 1990). But the major problem of usage of the glu-Ara-C in cancer treatment is the release time of the drug. The prolonged release and inhibition of cytidine deaminase play an essential role in enhancement of the antitumor effect of Ara-C (Aoshima, Tsukagoshi, Sakurai, Oh-ishi, & Ishida, 1977; Kato, Saito, Fukushima, Takeda, & Hara, 1984; Onishi, Seno, Pithayanukul, & Nagai, 1991). Chitin can conjugate with glu-Ara-C (Chi-glu-Ara-C) to extend the release time of the drug. The antitumor effect of Chi-glu-Ara-C was investigated by intraperitoneal administration to mice intraperitoneally inoculated with P388 leukemia (Ichikawa, Onishi, Takahata, Machida, & Nagai, 1993).

Carboxymethyl chitin (CMC) is also used for drug delivery application. The hydrophobic anticancer drug 5FU was loaded into CMC nanoparticles via emulsion cross-linking method. Drug release studies showed that the CMC nanoparticles provided a controlled and sustained drug release at pH 6.8 (Manjusha et al., 2010). Moreover CMC is a promising biopolymer for cancer diagnosis application also. Manjusha et al. developed novel folic acid-conjugated CMC coordinated to manganese-doped zinc sulfide (ZnS:Mn) quantum dot (FA-CMCS-ZnS:Mn) nanoparticles. The system can be used for targeting, controlled drug delivery, and also imaging of cancer cells. The biocompatible FA-CMCS-ZnS:Mn was used on breast cancer cell line MCF-7 to study the imaging, specific targeting, and cytotoxicity of the drug-loaded nanoparticles. The results showed that the bright and stable luminescence of quantum dots can be used to image the drug carrier in cancer cells without affecting their metabolic activity and morphology (Manjusha et al., 2010).

For gene delivery to cancer cells, several polymers have been used as nonviral vectors (Aoki et al., 2001; Miyata et al., 2008; Vernejoul et al., 2002; You, Manickam, Zhou, & Oupický, 2007). Even if low solubility and transfection efficiency is limiting the usage of chitosan in gene therapy applications, chitosan is a promising candidate as a vector for gene delivery to cancer cells. Therefore, researchers have modified this polymer in order to get an effective transfection. For instance, Germershaus, Mao, Sitterberg, Bakowsky, & Kissel (2008), Kean, Roth, & Thanou (2005), and Thanou, Florea, Geldof, Junginger, & Borchard (2002) quaternize the chitosan. According to their researches on quaternized chitosan derivatives, properties such as their charge, solubility, plasmid interactions, and transfections were increased. In 2011, Safari et al. prepared N,N-diethyl N-methyl chitosan (DEMC) for gene delivery to human pancreatic cancer cells (Safari et al., 2011). According to their biological research and the mathematical modeling results, both showed that after DEMC transfection, cancer cell fluorescence intensity and size has been changed.

4. CONCLUSION

Although the surgical methods are still promising and widely accepted, treatments against defined cancer, nonsurgical treatments against cancer have also received much attention with an aim to reduce and eliminate complications after surgical treatments. Therefore overcoming side effect complication of the anticancer agent is the main scope of the cancer researchers. Recent studies of the chemical modification of chitin and chitosan are discussed from the viewpoint of biomedical applications because of their excellent biological properties such as biodegradation and biocompatibility in the human body. Such properties can be considered as valuable extensions of the use of chitin and its derivatives. These natural biological properties allow them to be valuable biomaterial for both anticancer therapy of human solid tumors and cancer diagnosis applications in various ways.

REFERENCES

Aoki, K., Furuhata, S., Hatanaka, K., Maeda, M., Remy, J. S., Behr, J. P., et al. (2001). Polyethylenimine-mediated gene transfer into pancreatic tumor dissemination in the murineperitoneal cavity. *Gene Therapy, 8*, 508–514.

Aoshima, M., Tsukagoshi, S., Sakurai, Y., Oh-ishi, J., & Ishida, T. (1976). Antitumor activities of newly synthesized N^4-acyl-1-beta-D-arabinofuranosylcytosine. *Cancer Research, 36*, 2726–2732.

Aoshima, M., Tsukagoshi, S., Sakurai, Y., Oh-ishi, J., & Ishida, T. (1977). N^4-behenoyl-1-beta-D-arabinofuranosylcytosine as a potential new antitumor agent. *Cancer Research, 37*, 2481–2486.

Artan, M., Karadeniz, F., Karagozlu, M. Z., Kim, M. M., & Kim, S. K. (2010). Anti-HIV-1 activity of low molecular weight sulfated chitooligosaccharides. *Carbohydrate Research, 345*, 656–662.

Cho, Y. I., Park, S., Jeong, S. Y., & Yoo, H. S. (2009). In vivo and in vitro anti-cancer activity of thermo-sensitive and photo-crosslinkable doxorubicin hydrogels composed of chitosan-doxorubicin conjugates. *European Journal of Pharmaceutics and Biopharmaceutics, 73*, 59–65.

Defaye, J., & Guillot, J. M. (1994). A convenient synthesis for anomeric 2-thioglucobioses, 2-thiokojibiose and 2-thiosophorose. *Carbohydrate Research, 253*, 185–194.

Felt, C., Buri, P., & Gurny, R. (1998). Chitosan: A unique polysaccharides for drug delivery. *Drug Development and Industrial Pharmacy, 24*, 979–993.

Germershaus, O., Mao, S., Sitterberg, J., Bakowsky, U., & Kissel, T. (2008). Gene delivery using chitosan, trimethyl chitosan or polyethyleneglycol-graft-trimethyl chitosan block copolymers: Establishment of structure–activity relationships in vitro. *Journal of Controlled Release, 125*, 145–154.

Gingras, M., Paradis, I., & Berthod, F. (2003). Nerve regeneration in a collagen-chitosan tissue-engineered skin transplanted on nude mice. *Biomaterials, 24*, 1653–1661.

Hahn, M. G. (1996). Microbial elicitors and their receptors in plants. *Annual Review of Phytopathology, 34*, 387–412.

Han, H. D., Song, C. K., Park, Y. S., Noh, K. H., Kim, J. H., Hwang, T. W., et al. (2008). A chitosan hydrogel-based cancer drug delivery system exhibits synergistic antitumor effects by combining with a vaccinia viral vaccine. *International Journal of Pharmaceutics, 350*, 27–34.

Harish Prashanth, K. V., & Tharanathan, R. N. (2005). Depolymerized products of chitosan as potent inhibitors of tumor-induced angiogenesis. *Biochimica et Biophysica Acta, 11*, 1722–1729.

Hasegawa, M., Yagi, K., Iwakawa, S., & Hirai, M. (2001). Chitosan induces apoptosis via caspase-3 activation in bladder tumor cells. *Japanese Journal of Cancer Research, 92*, 459–466.

Horowitz, S. T., Roseman, S., & Blumenthal, H. J. (1957). The preparation of glucosamine oligosaccharides. I. Separation. *Journal of American Chemical Society, 79*, 5046–5048.

Huang, R., Mendis, E., Rajapakse, N., & Kim, S. K. (2006). Strong electronic charge as an important factor for anticancer activity of chitooligosaccharides (COS). *Life Sciences, 78*, 2399–2408.

Ichikawa, H., Onishi, H., Takahata, T., Machida, Y., & Nagai, T. (1993). Evaluation of the conjugate between N^4-(4-carboxybutyryl)1-β-D-arabinofuranosylcytosine. *Drug Design and Discovery, 10*, 343–353.

Izume, M., & Ohtakara, A. (1987). Preparation of D-glucosamine oligosaccharides by the enzymatic hydrolysis of chitosan (biological chemistry). *Bioscience, Biotechnology, and Biochemistry, 51*, 1189–1191.

Jeon, Y. J., Park, P. J., & Kim, S. K. (2001). Antimicrobial effect of chitooligosaccharides produced by bioreactor. *Carbohydrate Polymers, 44*, 71–76.

Karadeniz, F., Artan, M., Kim, M. M., & Kim, S. K. (2008). Prevention of cell damage on pancreatic beta cells by chitooligosaccharides. *Journal of Biotechnology, 136*, 539–540.

Karagozlu, M. Z., Karadeniz, F., Kong, C. S., & Kim, S. K. (2012). Aminoethylated chitooligomers and their apoptotic activity on AGS human cancer cells. *Carbohydrate Polymers, 87*, 1383–1389.

Kato, Y., Saito, M., Fukushima, H., Takeda, Y., & Hara, T. (1984). Antitumor activity of 1-beta-D-arabinofuranosylcytosine conjugated with polyglutamic acid and its derivative. *Cancer Research, 44*, 25–30.

Kean, T., Roth, S., & Thanou, M. (2005). Trimethylated chitosans as non-viral gene delivery vectors: Cytotoxicity and transfection efficiency. *Journal of Controlled Release, 103*, 643–653.

Khnor, E., & Lim, L. (2003). Implanted applications of chitin and chitosan. *Biomaterials, 24*, 2339–2349.

Kong, C. S., Kim, J. A., Ahn, B., Byun, H. G., & Kim, S. K. (2010). Carboxymethylations of chitosan and chitin inhibit MMP expression and ROS scavenging in human fibrosarcoma cells. *Process Biochemistry, 45*, 179–186.

Kweon, D. K., Song, S. B., & Park, Y. Y. (2003). Preparation of water-soluble chitosan/heparin complex and its application as wound healing accelerator. *Biomaterials, 24*, 1595–1601.

Maeda, Y., & Kimura, Y. (2004). Antitumor effects of various low-molecular weight chitosans are due to increased natural killer activity intestinal intraphelial lymphocytes in sarcoma 180-bearing mice. *Nutrition and Cancer, 134*, 945–950.

Manjusha, E. M., Mohan, J. C., Manzoor, K., Nair, S. V., Tamura, H., & Jayakumar, R. (2010). Folate conjugated carboxymethyl chitosan-manganese doped zinc sulphide nanoparticles for targeted drug delivery and imaging of cancer cells. *Carbohydrate Polymers, 80*, 414–420.

Mao, J. S., Liu, H. F., Yin, Y. J., & Yao, K. D. (2003). The properties of chitosan-gelatin membranes and scaffolds modified with hyaluronic acid by different method. *Biomaterials, 24*, 1621–1629.

Miyata, K., Oba, M., Kano, M. R., Fukushima, S., Vachutinsky, Y., Han, M., et al. (2008). Polyplex micelles from triblockcopolymers composed of tandemly aligned segments with biocompatible, endosome escaping, and DNA-condensing functions for systemic gene delivery to pancreatic tumor tissue. *Pharmaceutical Research, 25*, 2924–2936.

Mori, T., Okumura, M., Matsuura, M., Ueno, K., Tokura, S., Okamoto, Y., et al. (1997). Effects of chitin and its derivatives on the proliferation and cytokine production of fibroblasts in vitro. *Biomaterials, 18*, 947–951.

Murata, J., Saiki, I., Matsuno, K., Tokura, S., & Azumo, I. (1990). Inhibition of tumor cell arrest in lungs by anti metastatic chitin heparimoid. *Japanese Journal of Cancer Research, 80*, 866–872.

Muzzarelli, R. A. A. (1997). Human enzymatic activities related to the therapeutic administration of chitin derivatives. *Cellular and Molecular Life Sciences, 53*, 131–140.

Nicol, S. (1991). Life after death for empty shells: Crustacean fisheries create a mountain of waste shells, made of a strong natural polymer, chitin. Now chemists are helping to put this waste to some surprising uses. *New Scientist, 1755*, 36–38.

Nishimura, K., Nishimura, S., Nishi, N., Saiki, I., Tokura, S., & Azuma, I. (1984). Immunological activity of chitin and its derivatives. *Vaccine, 2*, 93–99.

Nishiyama, Y., Yoshikawa, T., Ohara, N., Kurita, K., Hojo, K., Kamada, H., et al. (2000). A conjugate from a laminin-related peptide, Try-Ile-Gly-Ser-Arg, and chitosan: Efficient and regioselective conjugation and significant inhibitory activity against experimental cancer metastasis. *Journal of Chemical Society, Perkin Transactions 1*, 1161–1165.

Obara, K., Ishihara, M., Ozeki, Y., Ishizuka, T., Hayashi, T., Nakamura, S., et al. (2005). Controlled release of paclitaxel from photocrosslinked chitosan hydrogels and its subsequent effect on subcutaneous tumor growth in mice. *Journal of Controlled Release, 110*, 79–89.

Okamoto, Y., Minami, S., Matsuhashi, A., Sashiwa, H., Saimoto, H., Shigemasa, Y., et al. (1993). Polymeric N-acetyl-D-glucosamine (Chitin) induces histionic activation in dogs. *Journal of Veterinary Medical Science, 55*, 739–742.

Onishi, H., Pithayanukul, P., & Nagai, T. (1990). Antitumor characteristics of the conjugate of N^4-(4-carboxybutyryl)-ara-C with ethylenediamine-introduced dextran and its resistance to cytidine deaminase. *Drug Design and Delivery, 6*, 273–280.

Onishi, H., Seno, Y., Pithayanukul, P., & Nagai, T. (1991). Conjugate of ethylenediamine introduced dextran. Drug release profiles and further in vivo study of its antitumor effects. *Drug Design and Delivery, 7*, 139–145.

Ouchi, T., Inosaka, K., Banba, T., & Ohya, Y. (1992). *Design of chitin or chitosan/5-fluorouracil conjugate having antitumor activity*. United Kingdom: Elsevier Applied Science.

Park, P. J., Je, J. Y., & Kim, S. K. (2004). Free radical scavenging activities of differently deacetylated chitosans using an ESR spectrometer. *Carbohydrate Polymers, 55*, 17–22.

Qin, C., Du, Y., Xiao, L., Li, Z., & Gao, X. (2002). Enzymic preparation of water-soluble chitosan and their antitumor activity. *International Journal of Biological Macromolecules, 31*, 111–117.

Safari, S., Dorkoosh, F. A., Soleimani, M., Zarrintan, M. H., Akbari, H., Larijani, B., et al. (2011). N-diethylmethyl chitosan for gene delivery to pancreatic cancer cells and the relation between charge ratio and biologic properties of polyplexes via interpolations polynomial. *International Journal of Pharmaceutics, 420*, 350–357.

Sashiwa, H., Saimoto, H., Shigemasa, Y., Ogawa, R., & Tokura, S. (1990). Lysozyme susceptibility of partially deacetylated chitin. *International Journal of Biological Macromolecules, 90*, 295–296.

Sato, T., Ishii, T., & Okahata, Y. (2001). In vitro gene delivery mediated by chitosan. *Biomaterials, 22*, 2075–2080.

Seo, W. G., Pae, H. O., Kim, N. Y., Oh, G. S., Park, I. S., Kim, Y. H., et al. (2000). Synergistic cooperation between water-soluble chitosan oligomers and interferon-[gamma]

for induction of nitric oxide synthesis and tumoricidal activity in murine peritoneal macrophages. *Cancer Letters, 159,* 189–195.

Shigemasa, Y., Saito, K., Sashiwa, H., & Saimoto, H. (1994). Enzymatic degradation of chitins and partially deacetylated chitins. *International Journal of Biological Macromolecules, 16,* 43–49.

Sirica, A. E., & Woodman, R. J. (1971). Selective aggregation of L1210 leukemia cells by the polycation chitosan. *Journal of the National Cancer Institute, 47,* 377–388.

Suzuki, K., Mikami, T., Okawa, Y., Tokoro, A., Suzuki, S., & Suzuki, M. (1986). Antitumor effect of hexa-N-acetylchitohexaose and chitohexaose. *Carbohydrate Research, 151,* 403–408.

Tanigawa, T., Tanaka, Y., Sashiwa, H., Saimoto, H., & Shigemasa, Y. (1992). Various biological effects of chitin derivatives. In C. J. Brine, P. A. Sandford, & J. P. Zikakis (Eds.), *Advances in chitin and chitosan* (pp. 206–215). London: Elsevier.

Thanou, M., Florea, B. I., Geldof, M., Junginger, H. E., & Borchard, G. (2002). Quaternized chitosan oligomers as novel gene delivery vectors in epithelial cell lines. *Biomaterials, 23,* 153–159.

Tokoro, A., Tatewaki, N., Suzuki, K., Mikami, T., Suzuki, S., & Suzuki, M. (1988). Growth inhibitory effect of hexa-N-acetylchitohexaose and chitohexaose against Meth-A solid tumor. *Chemical & Pharmaceutical Bulletin, 36,* 784–790.

Tokura, S., Ueno, K., Miyazaki, S., & Nishi, N. (1997). Molecular weight dependent antimicrobial activity by chitosan. *Macromolecular Symposia, 120,* 1 9.

Tsukada, S., & Inoue, Y. (1981). Conformational properties of chito-oligosaccharides: titration, optical rotation, and carbon-13 NMR studies of chito-oligosaccharides. *Carbohydrate Research, 88,* 19–38.

Vernejoul, F., Faure, P., Benali, N., Calise, D., Tiraby, G., Pradayrol, L., et al. (2002). Antitumor effect of in vivo somatostatin receptor subtype 2 gene transfer in primary and metastatic pancreatic cancer models. *Cancer Research, 62,* 6124–6131.

Wang, Y. C., Lin, M. C., Wang, D. M., & Hsieh, H. J. (2003). Fabrication of a novel porous PGA-chitosan hybrid matrix for tissue engineering. *Biomaterials, 24,* 1047–1057.

You, Y. Z., Manickam, D. S., Zhou, Q. H., & Oupický, D. (2007). Reducible poly (2 dimethyl aminoethyl methacrylate): Synthesis, cytotoxicity, and gene delivery activity. *Journal of Controlled Release, 122,* 217–225.

INDEX

Note: Page numbers followed by "*f*" indicate figures and "*t*" indicate tables.

A

Acid hydrolysis, 99–100
Actinobacteria, 88–89
ADH. *See* Adipic dihydrazide (ADH)
Adipic dihydrazide (ADH), 142
Adipose-derived stem cells (ADSCs), 159–160
ADSCs. *See* Adipose-derived stem cells (ADSCs)
Alginates
  biological effects, 97–98
  brown algae, 96
  DMPO–OH, ESR spectra, 106–108, 107*f*
  enzymatic digestion, 101–102
  Fenton reaction, 106–107
  hydroxyl radical scavenging activity, 107–108, 107*f*
  MAP kinase inhibitors, 103–105
  nitric oxide (NO), 105
  oligosaccharides, 96–97
  physiological activities, 96
  preparation, 98–100
  RAW264.7 cells, TNF-α secretion, 100–101
  ROS, 106–107
  sodium, 98, 99*t*
  spin-trapping agent DMPO, 106–107
Alkyltrimethylammonium, 143–144
Anticancer therapy, fucoidans
  activated NK cells, 203–204
  apoptosis, 202–203
  caspase-3, 202–203
  cell proliferation, HCC, 201–202
  death receptor stimulation, 202–203
  DNA fragmentation, 203–204
  epidemiological studies, 200
  HCT-15 cells, 203–204
  LMWF, 200–201
  metastasis, 200
  mitochondrial depolarization, 202–203
  natural biodiversity, 200

  T-cell lymphoma, 201–202
  therapeutic strategies, 200
Anticoagulant activity, HS
  Cyprinia islandica and *Mactrus pussula*, 126–127
  factor Xa and anti-thrombin III, 128
  K. opima and *A. rhombea*, 132
  T. maxima and *P. viridis*, 129, 129*t*, 132
Anti-HIV activity, 117–118
Antihuman papillomavirus (anti-HPV) activity, 118
Antioxidant activities
  alginates (*see* Alginates)
  carrageenan, 120
Antiproliferation
  HA, 64, 71
  MCF-7 cells, 202–203
  PASM cell growth, 132
  serum concentration, 128
  SMC growth *in vitro*, 133
Anti-thrombin III (AT III), 48–49
Antitumor activity
  chitin (*see* Chitin)
  chitosan (*see* Chitosan)
AT III. *See* Anti-thrombin III (AT III)
Atom transfer radical polymerization (ATRP), 142, 144
ATRP. *See* Atom transfer radical polymerization (ATRP)

B

Biomedical applications, HA
  atherosclerosis, 148–149
  bacterial contamination, materials, 146
  biofilm formation, 146
  brushite cements, 151
  bulk gels/hydrogel particles, 147
  microneedle fabrication, 151
  nisin, 146
  OA, 149
  organic–inorganic materials, 150–151
  orthopedics, 150

Biomedical applications, HA (*Continued*)
  osteotransductive CPCs, 151
  PEG, 148
  photopolymerization, 148
  polymer conjugation, 148
  SF, 149
Blood coagulation, 117, 126–127
Bone TE
  bone marrow stromal cells, 154
  HA/CS bilayered scaffold, 154
  hADSCs, 154–155
  HA–EDA, 154
  titanium (Ti) and alloy, 153–154
Brain TE, 158–159

C

Calcium phosphate cements (CPCs)
  hydroxyapatite, 147
  osteotransductive, 151
Cancer and fucoidans
  anaplastic cell proliferation, 195–196
  angiogenesis, 204–206
  growth signaling molecules, 205–206, 207*f*
  marine natural products, 196
  metastasis, 204, 205–206
  natural products, 196
  phaeophyceae/brown algae, 196
  pharmaceutical and medical research, 196
Carbon nanotubes (CNTs), 168
Carboxymethylcellulose sodium (CMC-Na), 142
Carrageenans
  and agars, 114
  anticoagulants, 117
  antioxidant activity, 120
  antitumor activity, 120–121
  antiviral agents, 117–119
  cholesterol-lowering effects, 119
  classification, seaweeds, 113–114
  endogenous factor, 115
  food and technological applications, 121–122
  Greek symbols, 114
  hybrid (*see* Hybrid carrageenans)
  immunomodulatory activity, 119–120
  production methods, 115, 116*f*

Cartilage TE
  disadvantages, 156–157
  dual-syringe setup, 156
  electrospun mats, HA, 156
  GAGs, 156
  Gel–MA hydrogels, 155–156
  MSCs, 156
Chitin
  biological properties, 216
  biomaterial matrixes, 2
  biowastes, 2
  chemical process, 6
  crab and shrimp shells, 167
  degree of deacetylation and acetylation, 11–13
  as drug carriers, 220–221
  extraction process, 3–6
  fermentation, 6
  grinding and blending machine, 9–11, 11*f*
  international official standard methods, 2–3
  LMW and chito-oligosaccharides, 6–9
  marine crustacean sources, 3
  MMP inhibition, 219–220
  molecular weight determination, 13–14
  natural polysaccharide, 215–216
  NMR spectrometer, 11–13, 13*f*
  nomenclature system, 2–3
  ordered crystalline microfibril, 216
  physical appearance testing, 9
  plant growth stimulator, 2–3
  radical-induced oxidative damage, 219–220
  SCM–chitin, 217
  seafood processing industry, 3, 4*f*
  shells, crabs and shrimps, 3–6, 4*f*
  shrimp waste demineralization, 3–6
  sieves and mechanical shaker, 9–11, 12*f*
  turbidity determination, 11, 12*f*
Chitooligosaccharides (COS)
  degraded oligomers, 216–217
  depolymerization, chemical method, 7–8
  enzymatic method, depolymerization, 9, 10*t*
  fluorescence microscopic observations, 219
  intraperitoneal injection, 219
  nutraceuticals, 219
  T-cell proliferation, 219

Chitosan
   adsorption applications, 167
   antimicrobial sponge, 146–147
   biomaterial matrixes, 2
   biomedical and pharmaceutical industries, 216–217
   biowastes, 2
   chemical process, 6
   chitooligosaccharides (COS), 216–217
   CNTs, 168
   cytolytic T-lymphocytes, 217–219
   degree of deacetylation and acetylation, 11–13
   as drug carriers, 220–221
   enzymatic hydrolysis, 217–219
   extraction process, 3–6
   fermentation, 6
   gene delivery, 216–217
   grinding and blending machine, 9–11, 11f
   hexameric chitosan oligomer, 217–219
   "*in situ*" biodegradable gel, 153
   international official standard methods, 2–3
   laminin, 219
   LMWC and chito-oligosaccharides, 6–9
   lymphokines, 217–219
   marine crustacean sources, 3
   molecular weight determination, 13–14
   natural nontoxic heteropolysaccharide, 216–217
   NMR spectrometer, 11–13, 13f
   nomenclature system, 2–3
   physical appearance testing, 9
   plant growth stimulator, 2–3
   seafood processing industry, 3, 4f
   shells, crabs and shrimps, 3–6, 4f
   shrimp waste demineralization, 3–6
   sieves and mechanical shaker, 9–11, 12f
   turbidity determination, 11, 12f
Chondroitin sulfate (CS), 56–58
CNTs. *See* Carbon nanotubes (CNTs)
CPCs. *See* Calcium phosphate cements (CPCs)
Creutzfeldt–Jakob disease, 49–50
CS. *See* Chondroitin sulfate (CS)
Cyanobacteria
   Cyanothece sp., 87, 88
   habitats, 87
   physicochemical properties, 88
   production, 87

D

DDS. *See* Drug delivery systems (DDS)
Dermal TE, 159–160
Dermatan sulfate (DS), 56–58
DFU. *See* Diabetic foot ulcer (DFU)
Diabetic foot ulcer (DFU), 146–147
Drug delivery systems (DDS), 160
Drug release, HA
   advantages, 160–161
   biodegradable polymer matrices, 160
   BMP-2, 161–162
   DDS, 160
   description, 160
   drug-loaded multilayered system, 161
   ectopic bone-forming assays, 161–162
   oral delivery, 161
   therapeutic agent(s), 160
DS. *See* Dermatan sulfate (DS)

E

Ectopic bone-forming assays, 161–162
Electrophoretic deposition (EPD), 150–151
Embryonic stem cells (ESCs), 155
Emulsifiers, 90, 121, 182–183
β-Endoglucuronidase, 46–47
Enhanced oil recovery, 90
Enzymatic digestion, alginates
   gel-filtration analysis, 98–99, 101–102
   growth, bifidobacteria/plants, 102
   I-S and ULV-3, elution profiles, 101–102, 103f
   MAP kinase inhibitors, 103–104, 104f
   nitric oxide (NO), 105
   treatment duration, 101, 102f
EPD. *See* Electrophoretic deposition (EPD)
Erythrosin B, 147
ESCs. *See* Embryonic stem cells (ESCs)
Extracellular polysaccharides (EPSs)
   advantages, 81
   antitumor, antiviral and immunostimulant activities, 80, 89
   biosynthesis, 83
   carbon/nitrogen ratio, 83
   description, 80
   dextran and xanthan, 81

Extracellular polysaccharides (EPSs) (*Continued*)
  energy cost, 81–82
  enhanced oil recovery, 90
  gelling agent, 89
  heavy metal removal, 90
  hydrated biofilms, 82
  industrial applications, 80
  isolation, 83
  microorganisms, 83–89
  surfactants and emulsifiers, 90

F

FCSPs. *See* Fucose-containing sulfated polysaccharides (FCSPs)
Fenton reaction, 106–107
Fibrin gel, 158
Food and technological applications, carrageenans, 121–122
Fourier transform rheology (FTR), 30, 32f, 39–40
FTR. *See* Fourier transform rheology (FTR)
Fucoidans
  anticancer effect, 200–204
  and cancer, 200–206
  chemical composition, 197–198
  heparin-like molecule, 197–198
  MMPIs (*see* MMP inhibitors (MMPIs))
  seaweed polysaccharides, 196–200
  structure and function, 198–200, 199f
Fucose-containing sulfated polysaccharides (FCSPs)
  biological functions, 198–200
  mice implanted Sarcoma-180 cells, 204–205
  native and oversulfated, 200–201
  sulfated galactofucans, 198
  types, 204–205

G

GAGs. *See* Glycosaminoglycans (GAGs)
Galactans, MMPIs
  agaran, chemical structure unit, 184, 184f
  angiogenesis, 183–184
  chemical structure unit, carrageenan, 182–183, 183f
  histochemical assays, 183–184
Gamma ray irradiation (GM), 6–7, 142–143

Gelling agents, 89, 122
Gel properties, hybrid carrageenans
  and chemical structure, 38–39
  concentration scaling, 36–38
  cooling solution, 35
  FTR, 30, 32f
  ionic strength, 35–36
  LAOS stress, 30
  mechanical spectra, 28–30, 31f, 34f
  MicroDSC data, 33–35
  penetration tests, 33
  power law, 36–38
  SAOS, 28–30, 29f
  steady shear, 39–40
  stiff and brittle gels, 33–35
  strain-hardening character, 39–40
  temperature sweep, 31f, 35, 36f
  viscoelastic behavior, 33–35
Gene therapy
  HAALD–PAHy hydrogel, 162
  PArg and polysaccharide HA, 163
  skin protection, 162–163
D-Glucopyranosyluronic acid, 46
Glycosaminoglycans (GAGs)
  blood coagulation, 126–127
  bovine eyes, 138
  cationic resin, 127–128
  cell growth, 130
  chondroitin sulfate (CS), 156
  collagen biofunctionality, 156
  factor Xa, 128
  heparin, 46
  HS, 129–130
  hyaluronic acid (HyA), 156
  natural macromolecules, 139
  P. viridis and *T. maxima*, 129–130, 133
  sulfated distribution, 51f
  tissues, 140–141
  types, 156
  vertebrates and invertebrates, 62–63, 126–127
GM. *See* Gamma ray irradiation (GM)

H

HA. *See* Hyaluronic acid (HA)
Halloysite nanotubes (HNTs), 150–151
Heart TE

Index

Alg and HyA, 157–158
fibrin gel, 158
vascular grafts, 157
Heparan sulfate (HS). *See also*
    Low-molecular-weight
    heparins/HS (LMWH/HS)
  animal sources, 52
  anticoagulant activity, 128
  antiproliferative activity, 128
  chemical structure, 52
  endothelial cells, 126, 133
  GAGs (*see* Glycosaminoglycans (GAGs))
  glucosaminido-iduronic acid linkages, 53
  ^{1}H-NMR, 128, 130
  isolation, 127–128
  mollusks, 125–126
  N-acetyl-D-glucosamine and
    D-glucuronic acid, 125–126
  sequences, 52
  structure, disaccharide units, 126, 127*f*
  T. maxima and *P. viridis*, 129, 129*t*
Heparin. *See also* Low-molecular-weight
    heparins/HS (LMWH/HS)
  anticoagulant activity, 47–49
  antithrombotic drug, 47
  biological actions, 47–48
  clinical anticoagulant, 47
  disaccharide units, 46
  GAG, 46
  gastrointestinal tract, 47–48
  genomics and sequencing, 58–59
  and HS (*see* Heparan sulfate (HS))
  marine sources, 50–52
  peptidoglycan, 46–47
  serum-plasma test, 47
  structure, 46*f*
  terrestrial sources, 49–50
Herpes simplex virus (HSV), 118
Heteropolysaccharides, 46, 139
High-molecular-weight (HMW), 138–139, 187
^{1}H-NMR spectroscopy, HS
  description, 128
  Perna viridis, 130, 131*f*
  Tridacna maxima, 130, 131*f*
HNTs. *See* Halloysite nanotubes (HNTs)
HS. *See* Heparan sulfate (HS)
HSV. *See* Herpes simplex virus (HSV)

Hyaluronic acid (HA)
  and ADH, 142
  amino acid-modified HA polymers, 142
  and ATRP, 142, 144
  biomedical applications, 146–151
  characterization and isolation, 71, 72*f*
  and CMC-Na, 142
  compositional analysis, 70
  description, 62
  and DPPH, 145
  drug delivery applications, 160–162
  and ECMs, 138–139
  electrophoretic analysis, 68
  environmental applications, 167
  enzyme digestion method, 66
  and GAG distribution, 62–63, 63*f*
  gamma irradiation, 145
  gene delivery applications, 162–163
  grafting techniques, 142
  and HA-DTPH, 144–145
  and HMWHA, 142–143
  human umbilical cord, 63–64
  hydrogels, 165
  and LBL technique, 143
  lipoic and lipohyal, 143
  LMWHA-1 and LMWHA-2, 145
  macroinitiators, 144
  market value, 71
  microbial production, 67
  molecular weight and viscosity
    determination, 70–71
  organic solvents and sodium acetate,
    66–67
  polyacid coupling, 143–144
  polyelectrolyte coupling, 143–144
  polysaccharides complex, 64
  properties, 141
  sensors, 168
  spectroscopic investigation, 68–69
  structure, 64, 66*f*, 139–141, 139*f*
  sulfated polysaccharides, 63
  supplementary methods, 67
  targeted drug delivery, 163–165
  TE applications, 152–160
  terrestrial ecosystem, 62–64, 65*t*
  tumor treatment, 166–167
Hybrid carrageenans
  algal material, 23–24

Hybrid carrageenans (*Continued*)
  alkali pretreatment, 26
  dark cultivation, 26–28
  description, 18
  extraction process, 23–24, 24f
  gel properties (*see* Gel properties, hybrid carrageenans)
  iota-carrageenan (I), 18
  kappa-carrageenan (K), 18
  KIMN, 18
  macromolecular structure, 21–23
  natural resource, 40
  postharvest storage, 25
  season variability, 24–25
  seaweeds chemistry, 19–21
Hydrogels, 165, 220

I

L-Idopyranosyluronic acid, 46
Immunomodulatory activity, 119–120
Interpenetrating polymeric network (IPN), 154–155
Intrinsic/extrinsic pathways, 49f
In vitro antioxidant assay, 145
Iota-carrageenan (I). *See* Hybrid carrageenans
IPN. *See* Interpenetrating polymeric network (IPN)

K

Kappa-carrageenan (K). *See* Hybrid carrageenans
Kappa/iota/mu/nu-hybrid carrageenan (KIMN). *See* Hybrid carrageenans

L

LAOS. *See* Large amplitude sinusoidal shear (LAOS)
Large amplitude sinusoidal shear (LAOS), 30, 32f, 39–40
Layer-by-layer (LBL) technique, 143
LMWF. *See* Low molecular weight fucoidan (LMWF)
LMWHA. *See* Low-molecular-weight hyaluronic acid (LMWHA)
LMWH/HS. *See* Low-molecular-weight heparins/HS (LMWH/HS)
Low-molecular-weight chitosan (LMWC)
  antitumor activity, 217–219
  depolymerization, chemical method, 7–8
  enzymatic method, depolymerization, 9, 10t
Low molecular weight fucoidan (LMWF)
  anticancer therapy, 206–207
  anti-inflammatory properties, 198–200
  balloon injury, thoracic aorta, 184–185
  biological actions, 198
  bone noncollagenous matrix, 198–200
  Ca^{2+} homeostasis, 203
  ER-positive MCF-7 cells, 203
  hind limb ischemia, rat model, 187
  mitochondrial-mediated pathway, 200–201
  tissue-rebuilding parameters, 198–200
Low-molecular-weight heparins/HS (LMWH/HS)
  anticoagulant activity, 58–59
  antithrombotic agents, 53–54
  chemical synthesis, 54–55
  chromatography separation, 56–58
  enzymatic synthesis, 55–56
  heparin species, 54
  marine products, 58
  materialization, 58–59
  molecular components, 53
  pharmacological functions, 53
  protein and ligand interactions array, 53
  United States, 54
Low-molecular-weight hyaluronic acid (LMWHA), 142–143, 145, 163
Lung TE, 152–153

M

Macroinitiators, HA, 144
MAP kinase inhibitors. *See* Mitogen-activated protein (MAP) kinase inhibitors
Marine bacteria, EPSs
  Alteromonas infernus sp., 86
  Bacillus licheniformis, 86
  chemical composition and molecular weight, 84
  Edwardsiella tarda, 86
  Enterobacter cloaceae (AK-I-MB-71a), 85
  haloalkalophilic *Bacillus* sp., 85
  heteropolysaccharide (PEP), 86–87

Pseudomonas sp. strain NCMB 2021, 84–85
Rhodococcus rhodochrous S-2, 85
16S rDNA sequences, *Pseudoalteromonas*, 83–84
Matrix metalloproteinases (MMPs)
  cell matrix adhesion, 179–180
  classes, 178, 178*f*
  inflammatory and immune responses, 179–180
  inhibition, chitin, 219–220
  intercellular regulation, 179–180
  "minimal domain", 178–179
  organ development, 178
  pathological processes, 179–180
  physiological processes, 178
  radical-induced oxidative damage, 219–220
  remodeling process, 178
  structure, 178–179, 179*f*
  tissue remodeling, 178
  tumor metastasis, 204
  zinc-dependent endopeptidases, 178
Mesenchymal stem cells (MSCs), 155
Mitogen-activated protein (MAP) kinase inhibitors, 103–105
MMP inhibitors (MMPIs)
  A549 lung cancer cells, 185
  astacin, 185–186
  batimastat, 186
  brown algae, 181
  cell proliferation and metastasis, 187–188
  cell surface proteins, 180
  cellular signaling transduction pathways, 187
  chemical structure, 184–185, 185*f*
  commercial synthetic MMPIs, 180–181, 181*f*
  epithelial-to-mesenchymal transition, 180
  genomic instability, 180
  hind limb ischemia, rat model, 187
  human cancer management, 180–181
  HUVECs, 184–185
  inflammatory reaction, 185–186
  marine macro algae, 181
  MMP-associated cerebral ischemia, 187
  molecular weight, 187
  nuclear antigen immunostaining, 184–185
  phlorotannins and polysaccharides, 181
  PLA2 enzyme-mediated lymphatic system damage, 186
  prozymogens, 180
  radical depolymerization, HMW, 187
  SPs, 184–188
  suppression effect, 187–188
  synthetic drugs, 180–181
  therapeutic ability, 187
  TIMPs, 180
  tumor microenvironment, 180
  VSMC, 184–185
  zinc peptidases, 185–186
MSCs. *See* Mesenchymal stem cells (MSCs)

N

N-acetyl/*N*-sulfate glucosaminido–glucuronic acid, 53
Nisin, 146
Nuclear magnetic resonance (NMR) spectrometer
  chitin and chitosan, 11–13, 13*f*
  FTIR methods, 13
  proton spectra, 133–134

P

PDI. *See* Polydispersity index (PDI)
Pentose phosphate pathway, 140–141
PEO. *See* Poly(ethylene oxide) (PEO)
Poly(ethylene oxide) (PEO), 144–145, 156
Poly(L-lactic acid) (PLLA), 156–157
Polydispersity index (PDI), 142
Polyelectrolyte multilayers, 143
Polysaccharide chains, 139, 140
Production methods, carrageenan, 115, 116*f*

R

Reactive oxygen species (ROS), 106–107, 149–150
Red seaweeds. *See* Carrageenans
Rheological characterization, HA, 141

S

SAOS. *See* Small amplitude oscillatory shear (SAOS)
Seaweed chemistry

Seaweed chemistry (*Continued*)
  disaccharide units, 19–21, 20*f*
  Gigartinales, 19–21
  M. stellatus and *Chondrus crispus*, DRIFT spectra, 19–21, 21*f*
  Portuguese coast, 19–21
  spectroscopic method, 19–21
Seaweed polysaccharides
  bioactive sulfated, 196–197
  fucans and alginic acid derivatives, 196–197
  fucoidans, 197–200
  human consumption, 196–197
  pharmacology and biochemistry, 196–197
Sensors, HA, 168
SF. *See* Synovial fluid (SF)
Small amplitude oscillatory shear (SAOS), 28–30, 29*f*
Sodium alginates, 98, 99*t*
SPs. *See* Sulfated polysaccharides (SPs)
Stem cells, TE, 155
Sulfated polysaccharides (SPs)
  age technologies, 188–189
  algae, marine sources, 182
  fucoidans, 184–188
  galactans, 182–184
  inflammatory responses, 188–189
  molecular size, 182
  Phaeophyta, fucans, 182
  Rhodophyta, galactans, 182–184
Synovial fluid (SF)
  and articular cartilage, 149
  and ECM, 163–164
  lubricating properties, 141
Synthetic drugs
  batimastat (BB-94), 180–181
  marimastat (BB-2516), 180–181

T

Targeted drug delivery, HA
  DOX, 164
  ECM and SF, 163–164
  HA–PLL conjugate, 164–165
  moieties, 163–164
  PEGylation, 163–164
  siRNA, 164–165
Tetraalkylammonium, 143–144
Tissue engineering (TE)
  bone, 153–155
  brain, 158–159
  cartilage, 155–157
  chitosan, 152
  collagen type I, 152
  dermal, 159–160
  description, 152
  heart, 157–158
  lung, 152–153
  stem cells, 155
TNF-α-inducing activities, alginates
  enzymatic digestion, 101–102
  MAP kinase inhibitors, 103–105
  molecular sizes and M/G ratios, RAW264.7 cells, 98, 99*t*, 100–101
Toothpastes, 121–122
Tumor treatment, HA, 166–167

U

UFH. *See* Unfractionated heparin (UFH)
Unfractionated heparin (UFH)
  vs. antithrombotic activity, 53–54
  depolymerization, chemical/enzymatic process, 56
  and LMWH, 53–54
  marine species, 55

V

Vascular smooth muscle cell (VSMC), 148–149, 184–185

Z

Zinc-dependent endopeptidases, 178, 185–186

CPI Antony Rowe
Eastbourne, UK
July 31, 2014